Lecture Notes in Business Information Processing 514

Series Editors

LNBIP reports state-of-the-art results in areas related to business information systems and industrial application software development – timely, at a high level, and in both printed and electronic form.

The type of material published includes

- Proceedings (published in time for the respective event)
- Postproceedings (consisting of thoroughly revised and/or extended final papers)
- Other edited monographs (such as, for example, project reports or invited volumes)
- Tutorials (coherently integrated collections of lectures given at advanced courses, seminars, schools, etc.)
- Award-winning or exceptional theses

LNBIP is abstracted/indexed in DBLP, EI and Scopus. LNBIP volumes are also submitted for the inclusion in ISI Proceedings.

João Araújo · Jose Luis de la Vara ·
Maribel Yasmina Santos · Saïd Assar
Editors

Research Challenges in Information Science

18th International Conference, RCIS 2024
Guimarães, Portugal, May 14–17, 2024
Proceedings, Part II

Editors
João Araújo
NOVA University Lisbon
Caparica, Portugal

Jose Luis de la Vara
University of Castilla La Mancha
Albacete, Albacete, Spain

Maribel Yasmina Santos
University of Minho
Guimarães, Portugal

Saïd Assar
Institut Mines-Télécom Business School
Evry, France

ISSN 1865-1348 ISSN 1865-1356 (electronic)
Lecture Notes in Business Information Processing
ISBN 978-3-031-59467-0 ISBN 978-3-031-59468-7 (eBook)
https://doi.org/10.1007/978-3-031-59468-7

This Springer imprint is published by the registered company Springer Nature Switzerland AG
The registered company address is: Gewerbestrasse 11, 6330 Cham, Switzerland

Paper in this product is recyclable.

Preface

This volume of the Lecture Notes in Business Information Processing series contains the proceedings of the 18th International Conference on Research Challenges in Information Science, RCIS 2024, held in Guimarães, Portugal, during May 14–17, 2024. Guimarães is a scenic historical city located in the north of the country and considered the cradle of the Portuguese nation.

The scope of RCIS covers the thematic areas of information systems and their engineering, user-oriented approaches, data and information management, enterprise management and engineering, domain-specific information systems engineering, data science, information infrastructures, and reflective research and practice. RCIS 2024 focused on the special theme "Information Science: Evolution or Revolution?".

The 25 full papers presented in the first volume were carefully reviewed and selected from a total of 79 submissions to the main conference. Out of all the submissions, three were desk rejected because the Program Co-chairs found them to be outside the scope of the conference. All the remaining submissions were single-blind reviewed by at least three Program Committee (PC) members. That was followed by a discussion period moderated by Program Board (PB) members. A PB meeting was also held in Lisbon to discuss the final selection of papers for the conference program. The acceptance rate was 32%. The authors received recommendations and meta-reviews to be contemplated in the camera-ready versions.

The second volume includes 12 Forum papers and five Doctoral Consortium papers. The Forum track received 14 dedicated submissions, out of which seven were accepted. Six submissions to the main conference were recommended for presentation at the Forum. The Doctoral Consortium received seven submissions, of which five were accepted.

The contributions in the first volume have been organized in the following topical sections: Data and Information Management, Conceptual Modelling and Ontologies, Requirements and Architecture, Business Process Management, Data and Process Science, Security, Sustainability, and Evaluation and Experience Studies; the second volume contains the Forum papers, Doctoral Consortium papers, and Tutorials.

The conference program started with the workshops. The main conference included sessions on keynotes, research papers, tutorials, the Forum, the Doctoral Consortium, research projects, and journal-first presentations. The three invited keynote presentations were: "Information science research with large language models: between science and fiction" by Fabiano Dalpiaz, University of Utrecht, The Netherlands; "The power of Information Systems shaping the future of the Automotive Industry", by Carlos Ribas, Bosch, Portugal; and "BPM in the Era of AI and Generative AI: Opportunities and Challenges" by Barbara Weber, University of St. Gallen (HSG), Switzerland. The four accepted tutorials addressed relevant and well-timed topics at the core interest of the RCIS community.

We want to thank all authors who submitted their work to RCIS 2024, and also the PC and PB members for their hard work in reviewing and discussing the submitted papers. Finally, we want to thank all the Organization Committee members and the student volunteers for their valuable assistance.

May 2024

João Araújo
Jose Luis de la Vara
Maribel Yasmina Santos
Saïd Assar

Organization

Conference Chairs

General Chairs

Maribel Yasmina Santos	University of Minho, Portugal
Saïd Assar	Institut Mines-Télécom Business School, France

Program Chairs

João Araújo	NOVA University Lisbon, Portugal
Jose Luis de la Vara	University of Castilla-La Mancha, Spain

Local Organizing Team

Carina Andrade	University of Minho, Portugal
Victor Barros	University of Minho, Portugal
Pedro Guimarães	CCG/ZGDV, Portugal
Paula Monteiro	CCG/ZGDV, Portugal
Isabel Ramos	University of Minho, Portugal
António Vieira	University of Minho, Portugal

Doctoral Consortium Chairs

Selmin Nurcan	Université Paris 1 Panthéon-Sorbonne, France
Jaelson Castro	Universidade Federal de Pernambuco, Brazil

Forum Chairs

Dominik Bork	TU Wien, Austria
Jānis Grabis	Riga Technical University, Latvia

Workshop Chairs

Jean-Michel Bruel	Toulouse University, France
Nelly Condori-Fernandez	Universidad Santiago de Compostela, Spain

Tutorial Chairs

Ana Moreira	NOVA University Lisbon, Portugal
Renata Guizzardi	University of Twente, The Netherlands

Research Projects Chairs

Dimitris Karagiannis	University of Vienna, Austria
Tiago Prince Sales	University of Twente, The Netherlands
Camille Salinesi	Université Paris 1 Panthéon-Sorbonne, France

Journal First Chairs

Angelo Susi	Fondazione Bruno Kessler, Italy
Jolita Ralyté	University of Geneva, Switzerland

Proceedings Chairs

Miguel Goulão	NOVA University Lisbon, Portugal
Oliver Karras	Leibniz ISCT, Germany

Publicity Chairs

Ana León	Universidad Politécnica de Valencia, Spain
Carla Silva	Universidade Federal de Pernambuco, Brazil
Isabel Sofia Brito	Polytechnic Institute of Beja, Portugal

Program Board

Saïd Assar	Institut Mines-Télécom Business School, France
Marko Bajec	University of Ljubljana, Slovenia
Xavier Franch	Universitat Politècnica de Catalunya, Spain
Renata Guizzardi	University of Twente, The Netherlands
Evangelia Kavakli	University of the Aegean, Greece
Pericles Loucopoulos	Institute of Digital Innovation and Research, UK
Haralambos Mouratidis	University of Essex, UK
Selmin Nurcan	Université Paris 1 Panthéon-Sorbonne, France
Oscar Pastor	Universidad Politécnica de Valencia, Spain
Jolita Ralyté	University of Geneva, Switzerland
Maribel Yasmina Santos	University of Minho, Portugal
Jelena Zdravkovic	Stockholm University, Sweden

Program Committee

Ademar Aguiar	University of Porto, Portugal
Nour Ali	Brunel University, UK
Raian Ali	Hamad Bin Khalifa University, Qatar
Jose María Alvarez Rodríguez	Carlos III University of Madrid, Spain
Carina Alves	Universidade Federal de Pernambuco, Brazil
Vasco Amaral	NOVA University Lisbon, Portugal
Claudia P. Ayala	Universitat Politècnica de Catalunya, Spain
Clara Ayora	University of Castilla-La Mancha, Spain
Fatma Başak Aydemir	Boğaziçi University, Turkey
Dominik Bork	TU Wien, Austria
Carlos Cetina	San Jorge University, Spain
Mario Cortes-Cornax	Université Grenoble Alpes, France
Maya Daneva	University of Twente, The Netherlands
Andrea Delgado	Universidad de la República, Uruguay
Rebecca Deneckere	Université Paris 1 Panthéon-Sorbonne, France
Chiara Di Francescomarino	University of Trento, Italy
Sophie Dupuy-Chessa	Université Grenoble Alpes, France
Sergio España	Utrecht University, The Netherlands
Robson Fidalgo	Universidade Federal de Pernambuco, Brazil
Hans-Georg Fill	University of Fribourg, Switzerland
Andrew Fish	University of Liverpool, UK
Agnès Front	Université Grenoble Alpes, France
Frederik Gailly	Ghent University, Belgium
Arturo García	University of Castilla-La Mancha, Spain
Ignacio García	University of Castilla-La Mancha, Spain
Sepideh Ghanavati	University of Maine, USA
Giovanni Giachetti	Universidad Andrés Bello, Chile
Cesar Gonzalez-Perez	Spanish National Research Council (CSIC), Spain
Jaap Gordijn	Vrije Universiteit Amsterdam, The Netherlands
Miguel Goulão	NOVA University Lisbon, Portugal
Giancarlo Guizzardi	Federal University of Espirito Santo (UFES), Brazil
Jennifer Horkoff	Chalmers University of Technology, Sweden
Felix Härer	University of Fribourg, Switzerland
Mirjana Ivanovic	University of Novi Sad, Serbia
Christos Kalloniatis	University of the Aegean, Greece
Oliver Karras	Leibniz ISCT, Germany
Manuele Kirsch Pinheiro	Université Paris 1 Panthéon-Sorbonne, France
Elena Kornyshova	Conservatoire National des Arts et Métiers, France

Tong Li	Beijing University of Technology, China
Grischa Liebel	Reykjavik University, Iceland
Lidia Lopez	Universitat Politècnica de Catalunya, Spain
Andrea Marrella	Sapienza University of Rome, Italy
Beatriz Marín	Universidad Politécnica de Valencia, Spain
Raimundas Matulevicius	University of Tartu, Estonia
Nikolay Mehandjiev	University of Manchester, UK
Giovanni Meroni	Technical University of Denmark, Denmark
Denisse Muñante	ENSIIE, France
Elena Navarro	University of Castilla-La Mancha, Spain
Kathia Oliveira	Université Polytechnique Hauts-de-France, France
Jose Ignacio Panach Navarrete	Universitat de València, Spain
Henderik A. Proper	TU Vienna, Austria
Manfred Reichert	University of Ulm, Germany
Patricia Rogetzer	University of Twente, The Netherlands
Marcela Ruiz	Zurich University of Applied Sciences, Switzerland
Alberto Sardinha	Universidade de Lisboa, Portugal
Rainer Schmidt	Munich University of Applied Sciences, Germany
Florence Sedes	I.R.I.T. Univ. Toulouse III Paul Sabatier, France
Sagar Sen	Simula Research Laboratory, Norway
Samira Si-Said Cherfi	Conservatoire National des Arts et Métiers, France
Carla Silva	Universidade Federal de Pernambuco, Brazil
Denis Silveira	Universidade Federal de Pernambuco, Brazil
Anthony Simonofski	Université de Namur, Belgium
Pnina Soffer	University of Haifa, Israel
Isabel Sofia	Polytechnic Institute of Beja, Portugal
Erick Stattner	University of the French West Indies, France
Monika Steidl	University of Innsbruck, Austria
Eric-Oluf Svee	Stockholm University, Sweden
Ernest Teniente	Universitat Politècnica de Catalunya, Spain
Olivier Teste	IRIT, France
Nicolas Travers	Pôle Universitaire Léonard de Vinci, France
Juan Trujillo	University of Alicante, Spain
Aggeliki Tsohou	Ionian University, Greece
Tanja E. J. Vos	Universidad Politécnica de Valencia, Spain
Yves Wautelet	Katholieke Universiteit Leuven, Belgium
Hans Weigand	Tilburg University, The Netherlands

Program Committee, Forum

Syed Juned Ali	TU Wien, Austria
Clara Ayora	Universidad de Castilla-La Mancha, Spain
Judith Barrios Albornoz	University of Los Andes, Colombia
Cinzia Cappiello	Politecnico di Milano, Italy
Istvan David	McMaster University, Canada
Victoria Döller	University of Vienna, Austria
Anne Gutschmidt	University of Rostock, Germany
Simon Hacks	Stockholm University, Sweden
Abdelaziz Khadraoui	University of Geneva, Switzerland
Manuele Kirsch Pinheiro	Université Paris 1 Panthéon-Sorbonne, France
Elena Kornyshova	Conservatoire National des Arts et Métiers, France
Georgios Koutsopoulos	Stockholm University, Sweden
Emanuele Laurenzi	FHNW School of Business, Switzerland
Dejan Lavbič	University of Ljubljana, Slovenia
Beatriz Marín	Universidad Politécnica de Valencia, Spain
Patricia Martin-Rodilla	Spanish National Research Council, Spain
Giovanni Meroni	Technical University of Denmark, Denmark
João Moura-Pires	NOVA University Lisbon, Portugal
Mark Mulder	TEEC2 BV, The Netherlands
Christoforos Ntantogian	Ionian University, Greece
Michalis Pavlidis	University of Brighton, UK
Francisca Pérez	Universidad San Jorge, Spain
Iris Reinhartz-Berger	University of Haifa, Israel
Ben Roelens	Ghent University, Belgium
Marcela Ruiz	Zurich University of Applied Sciences, Switzerland
Natalia Stathakarou	Karolinska Institutet, Sweden
Gianluigi Viscusi	Linköping University, Sweden
Manuel Wimmer	Johannes Kepler University Linz, Austria

Program Committee, Doctoral Consortium

Ademar Aguiar	University of Porto, Portugal
José Borbinha	Universidade de Lisboa, Portugal
Nelly Condori-Fernández	Universidad Santiago de Compostela, Spain
Maya Daneva	University of Twente, The Netherlands
João Faria	University of Porto, Portugal
Xavier Franch	Universitat Politècnica de Catalunya, Spain

Renata Guizzardi	University of Twente, The Netherlands
Hugo Jonker	Open University of the Netherlands, The Netherlands
Manuele Kirsch Pinheiro	Université Paris 1 Panthéon-Sorbonne, France
Beatriz Marín	Universidad Politécnica de Valencia, Spain
Oscar Pastor	Universidad Politécnica de Valencia, Spain
Fethi Rabhi	University of New South Wales, Australia
Jolita Ralyté	University of Geneva, Switzerland
Rogier Van de Wetering	Open University of the Netherlands, The Netherlands

Program Committee, Research Projects

Alessandra Bagnato	Softeam, France
Xavier Boucher	École nationale supérieure des mines de Saint-Étienne, France
Robert Buchmann	Babes-Bolyai University, Romania
Nelly Condori-Fernández	Universidad Santiago de Compostela, Spain
Tolga Ensari	Arkansas Tech University, USA
Davide Fucci	Blekinge Tekniska Högskola, Sweden
Filippo Lanubile	University of Bari, Italy
Khaled Medini	École nationale supérieure des mines de Saint-Étienne, France
Marc Oriol Hilari	Universitat Politècnica de Catalunya, Spain
Dalila Tamzalit	University of Nantes, France
Wilfrid Utz	OMILAB NPO, Germany
Tanja E. J. Vos	Politécnica de Valencia, Spain
Robert Woitsch	BOC Group, Austria

Additional Reviewers

Renata Cruz
Nikolaos Marios Polymenakos
Maxwell Prybylo
Sara Haghighi
Vijanti Ramautar
Ioannis Paspatis
Roberto Sanchez Reolid
Katerina Soumelidou
Simone Agostinelli
Sara Haghighi
Stylianos Karagiannis
Simon Dechamps
Stylianos Karagiannis
Martin Eisenberg
Evita Roponena

Abstracts of Keynote Talks

Information Science Research with Large Language Models: Between Science and Fiction

Fabiano Dalpiaz

Utrecht University, The Netherlands
f.dalpiaz@uu.nl

Abstract. Large language models (LLMs) are in the spotlight. Laypeople are aware of and are using LLMs such as OpenAI's ChatGPT and Google's Gemini on a daily basis. While companies are exploring new business opportunities, researchers have gained access to an unprecedented scientific playground that allows for fast experimentation with limited resources and immediate results. In this talk, using concrete examples from requirements engineering, I am going to put forward several research opportunities that are enabled by the advent of LLMs. I will show how LLMs, as a key example of modern AI, unlock research topics that were deemed too challenging until recently. Then, I will critically discuss the perils that we face when it comes to planning, conducting, and reporting on credible research results following a rigorous scientific approach. This talk will stress the inherent tension between the exciting affordances offered by this new technology, which include the ability to generate non-factual outputs (fiction), and our role and societal responsibility as information scientists.

The Power of Information Systems Shaping the Future of the Automotive Industry

Carlos Ribas

Bosch, Portugal
carlos.ribas@pt.bosch.com

Abstract. The automotive industry is undergoing a seismic shift, fueled by the convergence of information systems and artificial intelligence. In this keynote, I will explore how these technologies are reshaping the whole automotive industry, from the idea to the innovative creation, development, and manufacturing through to the customer experience. Join me in my travel through data highways, decoding smart factories and steering towards intelligent, autonomous and safe vehicles as assistants and companions.

BPM in the Era of AI and Generative AI: Opportunities and Challenges

Barbara Weber

University of St. Gallen, Switzerland
barbara.weber@unisg.ch

Abstract. Artificial intelligence (AI) technologies, such as machine learning and natural language processing, empower Business Process Management (BPM) by enabling data-driven decision-making, predictive analytics, and automation. Leveraging AI algorithms, organizations can extract actionable insights from vast datasets, optimize processes, and enhance operational efficiency to boost productivity. Additionally, Generative AI introduces novel capabilities to BPM, facilitating creative problem-solving and innovation. Generative AI algorithms, exemplified by Large Language Models (LLMs), not only offer enhanced creativity but also excel in tasks such as text generation, context understanding, and natural language interaction. With generative AI algorithms, alternative solutions and scenarios can be generated, augmenting human creativity and driving organizational innovation. In this keynote presentation I will highlight how this impacts the field of BPM and discuss some of the challenges arising from that.

Contents – Part II

Forum Papers

Contents – Part I

Business Process Management

Data and Process Science

Security

Sustainability

Evaluation and Experience Studies

Forum Papers

Transfer Learning for Potato Leaf Disease Detection

Shahid Mohammad Ganie[1], K. Hemachandran[1], and Manjeet Rege[2(✉)]

[1] AI Research Centre, Department of Analytics, Woxsen University, Hyderabad 502345, India
[2] Department of Software Engineering and Data Science, University of St. Thomas, Saint Paul, MN, USA
rege@stthomas.edu

Abstract. Deep learning techniques have demonstrated significant potential in the agriculture sector to increase productivity, sustainability, and efficacy for farming practices. Potato is one of the world's primary staple foods, ranking as the fourth most consumed globally. Detecting potato leaf diseases in their early stages poses a challenge due to the diversity among crop species, variations in symptoms of crop diseases, and the influence of environmental factors. In this study, we implemented five transfer learning models including VGG16, Xception, DenseNet201, EfficientNetB0, and MobileNetV2 for a 3-class potato leaf classification and detection using a publicly available potato leaf disease dataset. Image preprocessing, data augmentation, and hyperparameter tuning are employed to improve the efficacy of the proposed model. The experimental evaluation shows that VGG16 gives the highest accuracy of 94.67%, precision of 95.00%, recall of 94.67%, and F1 Score of 94.66%. Our proposed novel model produced better results in comparison to similar studies and can be used in the agriculture industry for better decision-making for early detection and prediction of plant leaf diseases.

Keywords: Plant leaf disease detection · transfer learning · image processing · VGG16 · Xception · DenseNet201 · EfficientNetB0 · MobileNetV2

1 Introduction

Food security is gravely jeopardized by plant diseases, which aggravate the problems of food insecurity and malnutrition that persist in many regions of the world [1]. Plant infections can result in catastrophic crop losses, which can result in food shortages and price rises that disproportionately affect the disadvantaged [2]. Emerging plant diseases, such as those brought on by Phytophthora infestans and Ralstonia solanacearum, are especially alarming because they have the potential to quickly traverse international borders and catastrophically damage crops. Rising temperatures and altered weather patterns are giving plant diseases an additional chance to grow, further aggravating the issue [3].

Millions of people rely on potatoes as an essential source of nourishment and income, making them a key crop for ensuring global food security [4]. Potatoes are the fourth most significant food crop (after rice, maize, and wheat) with a global production of

J. Araújo et al. (Eds.): RCIS 2024, LNBIP 514, pp. 3–11, 2024.
https://doi.org/10.1007/978-3-031-59468-7_1

around 388 million metric tons in 2020 [5]. Potatoes are grown in over 100 countries, with China, India, and Russia being the top three producers [6]. Moreover, disadvantaged groups, like children and pregnant women, who might not have access to other nutrient-rich meals, rely heavily on potatoes as a source of nutrition [7]. Potatoes are a crucial commodity for smallholder farmers in developing nations because they are versatile crops that can be cultivated in a range of soil types and temperatures. Furthermore, potatoes are a significant source of income for farmers, especially in areas where other crops might not be as lucrative [8].

Potatoes are vulnerable to a variety of illnesses, which can have a substantial influence on both the yield and quality of the crop. One of the most destructive potato diseases, late blight disease is spread by the oomycete pathogen Phytophthora infestans and is a significant global barrier to the profitable production of potatoes. Globally, the expenses of managing late blight and yield decrease losses are expected to total more than $6 billion annually [9]. Worldwide potato cultivators are very concerned about late blight, which can result in total crop loss if it fails to be adequately managed [10]. Potato bacterial wilt, which is brought on by the bacterium Ralstonia solanacearum, is another significant disease that affects potatoes. In tropical and subtropical areas, bacterial wilt can cause potato plants to wilt, yellow, and eventually die. The potato virus Y, the potato leafroll virus, and the potato spindle tuber viroid are other ailments that harm potatoes. Efforts to address the threat of plant diseases to global food security primarily include improved disease surveillance and effective disease management strategies, as well as investment in research and development of disease-resistant crop varieties [11].

1.1 Contribution

This study demonstrates that transfer learning in smart agriculture improves the accuracy, efficiency, and reliability of disease prediction in the agriculture industry. The following are the key contributions of this work:

- Data augmentation is applied by applying shear_range, zoom_range, rotation_range, horizontal_flip, and vertical_flip.
- Five transfer learning models, viz. VGG16, Xception, DenseNet201, EfficientNetB0, and MobileNetV2 are tried with the processed dataset.
- The performance of the transfer learning models was validated using various evaluation metrics including Accuracy, Loss, Precision, Recall, and F1-score.
- The proposed model has shown significant accuracy and outperforms the other models compared to the state-of-the-art works.

2 Literature Review

Potato late blight is a devastating disease that poses a significant threat to potato crops worldwide [12]. Traditional methods of detecting the disease are time-consuming, expensive, and subjective. Early detection and accurate monitoring of late blight are crucial for timely disease management and minimizing yield losses [13]. In the last decade, there have been several advancements in artificial vision and image processing that have provided new opportunities for the development of automated systems for potato late blight detection.

One method for employing artificial vision to find potato late blight entails looking at pictures of leaves that have the disease's obvious symptoms [14]. To do this, infected leaves can be digitally photographed under regulated lighting circumstances, and then the regions of interest (ROI) corresponding to the diseased portions can be identified by segmentation and feature extraction. Additionally, classification tasks based on these spectral signatures have been carried out using machine learning methods like neural networks. For accuracy, speed, and objectivity, these technologies outperform conventional visual examination techniques by a wide margin. Bashish et al. [15] presented a method for detecting and classifying plant leaf diseases, including potato late blight, using image processing techniques and artificial neural networks (ANN). They achieved an overall accuracy of 93% in detecting and classifying the diseases. Mokhtar et al. [16] presented a method for detecting potato late blight using image processing techniques and a fuzzy inference system. They achieved an accuracy of 95% in detecting late blight in potato plants. Sladojevic et al. [17] presented a method for detecting plant diseases, including potato late blight, using deep learning techniques. They achieved an overall accuracy of 96.3% in detecting and classifying the diseases. Sholihati et al. [18] classified the Plant Village dataset using VGG16 and VGG19, and VGG16 had a 91% accuracy rate. They exploited pre-existing architectures without adding any original design elements, and their technological contribution was minimal. Sankaran et al. [19] reviewed the use of hyperspectral and multispectral imaging for the detection of plant diseases, including potato late blight. They discussed the advantages and limitations of these imaging techniques and highlighted the potential of these technologies for improving disease detection in agriculture. Other researchers have investigated alternate ways for detecting potato late blight, such as fluorescence imaging or thermal imaging, which show promise in boosting detection rates while lowering costs when compared to other standard diagnostic instruments, such as PCR testing or ELISA assays. Carvajalino et al. [20] predicted the percentage of late blight affectation using multispectral photos of potato crops obtained at the canopy level using a UAV carrying an affordable, lightweight RGB camera with a filter to capture the red edge and part of the NIR band (680–800 nm). They employed two types of artificial neural networks: MLPs and innovative deep learning convolutional neural networks (CNNs), as well as support vector regression (SVR) and Random Forests (RFs). Ghosh et al. [21] also explored several ways to build an effective classifier for potato leaf disease using RGB images. Their main objective was to develop a set of instructions for constructing an image repository that would result in efficient classification accuracy.

3 Proposed Methodology

The process of transfer learning is learning a new task from knowledge acquired in a previous one. We can use a pretrained model and save the time elapsed for training. Transfer learning performs exceptionally well, even on small datasets. In our research, the categorization of different classes of early blight, healthy, and late blight was achieved using 5 transfer learning models such as VGG16, Xception, DenseNet201, EfficientNetB0, and MobileNetV2. The workflow of the proposed model is presented in Fig. 1.

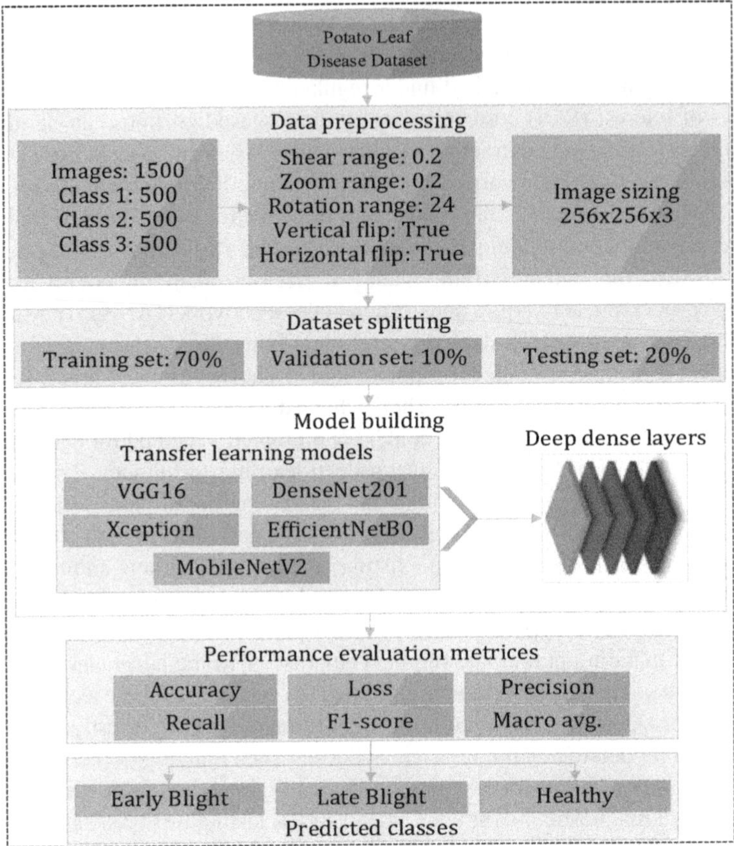

Fig. 1. Methodology for Prediction of Potato Leaf Disease

3.1 Dataset Selection and Description

A publicly available dataset[1] with JPG-format images of an average size of 256 × 256 pixels has been utilised to develop a transfer learning model for potato leaf disease. The dataset contains 1500 images and is separated into three categories: early blight, late blight, and healthy with the number of images in each category listed in Table 1. Also, Fig. 2 shows examples of the data samples.

Table 1. Detailed sample-wise distribution of the dataset.

Dataset	Total Images	Early blight	Late blight	Healthy
Potato Leaf Disease Dataset	1500	No. of Images = 500	No. of Images = 500	No. of Images = 500
		Class = 1	Class = 2	Class = 3
		Label = 0	Label = 1	Label = 2

[1] https://www.kaggle.com/datasets/muhammadardiputra/potato-leaf-disease-dataset.

Fig. 2. Samples images of each Class of the dataset.

4 Results and Discussion

In this section, the experimental details, including evaluation metrics of these transfer learning models, are covered. Also, the results are elaborately presented, and the performance of the proposed model is extensively assessed. Table 2 presents the class-wise values of different evaluation metrics including precision, recall, and F1-score.

Table 2. Class-wise values of precision, recall, and F1-score for considered models

Model	Class	Precision	Recall	F1-score
VGG16	Early Blight	0.98	0.98	0.98
	Healthy	0.88	0.97	0.92
	Late Blight	0.99	0.89	0.94
Xception	Early Blight	0.88	0.92	0.90
	Healthy	0.74	0.93	0.83
	Late Blight	0.94	0.66	0.78
DenseNet201	Early Blight	0.84	0.97	0.90
	Healthy	0.62	0.84	0.71
	Late Blight	1.00	0.49	0.66
EfficientNetB0	Early Blight	0.96	0.97	0.97
	Healthy	0.59	0.98	0.74
	Late Blight	1.00	0.33	0.50
MobileNetV2	Early Blight	0.83	0.97	0.89
	Healthy	0.81	0.86	0.83
	Late Blight	0.95	0.73	0.82

The training and testing accuracy of the considered transfer learning models is shown in Fig. 3. Among all the models, VGG 16 achieved the highest training and testing accuracy rate of 98.89% and 94.67% respectively. Xception performed worse during the training phase with an accuracy rate of 88.67% and EfficientNetB0 attained the lowest accuracy rate of 75.99% during the testing phase.

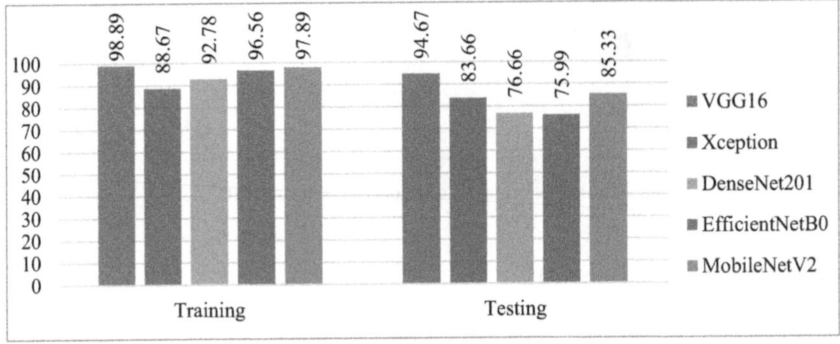

Fig. 3. Accuracy comparison of the transfer learning models

The training and testing loss of the considered transfer learning models is shown in Fig. 4. In terms of training and testing loss, DenseNet201 attained less training loss of 0.03%, while VGG 16 attained less testing loss of 0.14%.

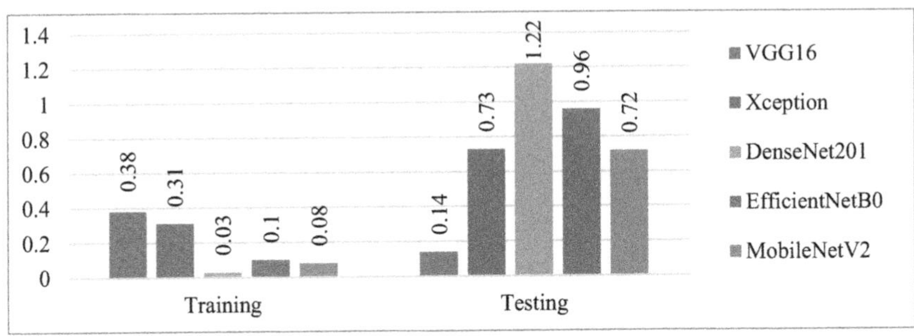

Fig. 4. Loss comparison of the transfer learning models

Figure 5 presents the macro average of the evaluation metrics including precision, recall, and f1-score. VGG16 outperformed other models and achieved the highest precision, recall, and f1-score at 0.95% respectively. DenseNet201 achieved the lowest precision rate of 0.82%, while EfficientNetB0 achieved the lowest recall and f1-score of 0.76%, and 0.73% respectively.

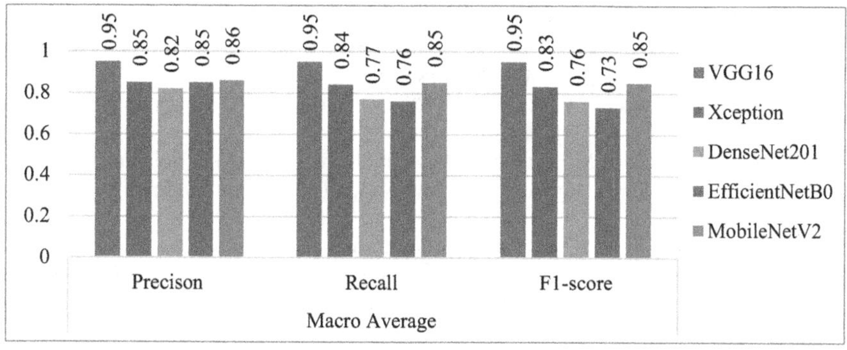

Fig. 5. Macro average of precision, recall, and F1-score of the transfer learning models

Overall, this study shows that transfer learning has the potential to improve how well image processing can be used to find leaf diseases across different crops. Using pre-trained deep neural networks can significantly reduce the need for large datasets and reduce training time, making them more accessible for smart agriculture applications. However, more research is needed to find the best architecture for transfer learning and the best fine-tuning strategy for the agriculture industry. Further studies can focus on improving the interpretability and generalisation of transfer learning models for real-world applications.

5 Conclusion

In this study, we developed a framework using transfer learning models for potato leaf disease detection. We primarily investigated the performance of all the considered transfer learning models including VGG16, Xception, DenseNet201, EfficientNetB0, and MobileNetV2 for potato leaf disease detection. The dataset was splitted into training, validation, and test sets for the model building process. The parameters used to train and test the models were then fine-tuned. To make the proposed model effective, we conducted data preprocessing and data augmentation. The performance of these models was evaluated using various metrics like accuracy, loss, precision, recall, f1-score, macro average, and weighted average. This demonstrates that the proposed framework can be used in the agriculture industry for smart farming. This work can be extended to detect leaf disease across multiple crops.

References

1. Nazarov, P.A., Baleev, D.N., Ivanova, M.I., Sokolova, L.M., Karakozova, M.V.: Infectious plant diseases: etiology, current status, problems and prospects in plant protection. Acta Naturae **12**(3), 46–59 (2020). https://doi.org/10.32607/actanaturae.11026

2. Rizzo, D.M., Lichtveld, M., Mazet, J.A.K., Togami, E., Miller, S.A.: Plant health and its effects on food safety and security in a One Health framework: four case studies. One Health Outlook **3**(1), 6 (2021). https://doi.org/10.1186/s42522-021-00038-7
3. Gautam, H.R., Bhardwaj, M.L., Kumar, R.: Climate change and its impact on plant diseases (2013)
4. Bajracharya, M., Sapkota, M.: Profitability and productivity of potato (Solanum tuberosum) in Baglung district, Nepal. Agric. Food Secur. **6**(1), 47 (2017). https://doi.org/10.1186/s40066-017-0125-5
5. Suo, H., et al.: Deep eutectic solvent-based ultrasonic-assisted extraction of phenolic compounds from different potato genotypes: comparison of free and bound phenolic profiles and antioxidant activity. Food Chem. **388**, 133058 (2022). https://doi.org/10.1016/j.foodchem.2022.133058
6. World Food and Agriculture – Statistical Yearbook 2021. FAO (2021). https://doi.org/10.4060/cb4477en
7. Kromann, L., Malchow-Møller, N., Skaksen, J.R., Sørensen, A.: Automation and productivity - a cross-country, cross-industry comparison. Ind. Corp. Chang. **29**(2), 265–287 (2020). https://doi.org/10.1093/icc/dtz039
8. Wasilewska-Nascimento, B., Boguszewska-Mańkowska, D., Zarzyńska, K.: Challenges in the production of high-quality seed potatoes (Solanum tuberosum L.) in the tropics and subtropics. Agronomy **10**(2) (2020). https://doi.org/10.3390/agronomy10020260
9. Strange, R.N., Scott, P.R.: Plant disease: a threat to global food security. Annu. Rev. Phytopathol. **43**(1), 83–116 (2005). https://doi.org/10.1146/annurev.phyto.43.113004.133839
10. Cao, W., et al.: Genome-wide identification and characterization of potato long non-coding RNAs associated with phytophthora infestans resistance. Front. Plant Sci. **12** (2021). https://doi.org/10.3389/fpls.2021.619062
11. Shoaib, M., et al.: An advanced deep learning models-based plant disease detection: a review of recent research. Front. Plant Sci. **14** (2023). https://doi.org/10.3389/fpls.2023.1158933
12. Arora, R.K.: Late blight disease of potato and its management. https://www.researchgate.net/publication/287301918
13. Anim-Ayeko, A.O., Schillaci, C., Lipani, A.: Automatic blight disease detection in potato (Solanum tuberosum L.) and tomato (Solanum lycopersicum, L. 1753) plants using deep learning. Smart Agric. Technol. **4**, 100178 (2023). https://doi.org/10.1016/j.atech.2023.100178
14. Griffel, L.M., Delparte, D., Whitworth, J., Bodily, P., Hartley, D.: Evaluation of artificial neural network performance for classification of potato plants infected with potato virus Y using spectral data on multiple varieties and genotypes. Smart Agric. Technol. **3**, 100101 (2023). https://doi.org/10.1016/j.atech.2022.100101
15. Al Bashish, D., Braik, M., Bani-Ahmad, S.: Detection and classification of leaf diseases using K-means-based segmentation and neural-networks-based classification. Inf. Technol. J. **10**(2), 267–275 (2011). https://doi.org/10.3923/itj.2011.267.275
16. Akther, J., Nayan, A.A., Harun-Or-roshid, M.: Potato leaves blight disease recognition and categorization using deep learning. Eng. J. **27**(9), 27–38 (2023). https://doi.org/10.4186/ej.2023.27.9.27
17. Sladojevic, S., Arsenovic, M., Anderla, A., Culibrk, D., Stefanovic, D.: Deep neural networks based recognition of plant diseases by leaf image classification. Comput. Intell. Neurosci. **2016** (2016). https://doi.org/10.1155/2016/3289801
18. Sholihati, R.A., Sulistijono, I.A., Risnumawan, A., Kusumawati, E.: Potato leaf disease classification using deep learning approach. In: IES 2020 - International Electronics Symposium: The Role of Autonomous and Intelligent Systems for Human Life and Comfort, Institute of Electrical and Electronics Engineers Inc., September 2020, pp. 392–397 (2020). https://doi.org/10.1109/IES50839.2020.9231784

19. Invasive alien plants in the forests of Asia and the Pacific new
20. Qi, C., et al.: In-field early disease recognition of potato late blight based on deep learning and proximal hyperspectral imaging
21. Chakraborty, K.K., Mukherjee, R., Chakroborty, C., Bora, K.: Automated recognition of optical image based potato leaf blight diseases using deep learning. Physiol. Mol. Plant Pathol. **117**, 101781 (2022). https://doi.org/10.1016/j.pmpp.2021.101781

Knowledge Graph Multilevel Abstraction: A Property Graph Reification Based Approach

Selsebil Benelhaj-Sghaier(✉), Annabelle Gillet, and Éric Leclercq

Laboratoire d'Informatique de Bourgogne - EA 7534, University of Burgundy, Dijon, France
Selsebil_Ben-El-Haj-Sghaier@etu.u-bourgogne.fr,
{annabelle.gillet,eric.leclercq}@u-bourgogne.fr

Abstract. Adding knowledge to data or information is an essential step for new information system, which need to break down silos in order to support a large diversity of applications. Conventional integration approaches have difficulties in meeting the flexibility required by new needs, mainly because they rely on rigid schema. Graph-based approaches, such as knowledge graphs, are promising, as they allow to use different graph models such as RDF or property graph models. However, they do not make it easy to describe complex relationships at different levels of abstraction. The reification process, already well-studied for RDF, is a promising solution to add new representation capabilities. This paper delves into the process of knowledge reification within the property graph model as a novel approach to enhance the expressivity of knowledge graph model by adding capabilities of representing complex relationships and multilevel abstractions. Based on the study of reification models for RDF, we formalize a new model for property graphs by generalizing the different reification techniques.

Keywords: Knowledge graph · Knowledge reification · Property graph · Multilevel knowledge representation

1 Introduction

As the variety and volume of data sources continue to increase, new challenges in handling data and knowledge emerge. Indeed, these data sources can be represented with various models, each possessing different characteristics and containing information of different nature [5]. In addition, data evolve rapidly, requiring data management tools with a high level of flexibility.

Furthermore, the multiplication of data-centric applications corresponding to new use cases require to break information system silos by defining multiple views of the data [1].

In this context, knowledge graphs [10,12] have gained increasing interest as a flexible solution to organize and represent knowledge in a structured format. They use a graph-based approach, where entities are represented as nodes and

J. Araújo et al. (Eds.): RCIS 2024, LNBIP 514, pp. 12–19, 2024.
https://doi.org/10.1007/978-3-031-59468-7_2

relationships between them are represented as edges. This structure makes it easy to store, retrieve, and analyze data, and it is particularly well-suited for a wide range of applications [22] such as search engines, recommendation systems, and medical knowledge integration.

Knowledge graphs allow to use different graph based data models, that can have different expressivity capabilities. For example, using a property graph model [2] instead of a RDF graph allows to add properties on relations. However, the expressivity of current graph models is still limited, and cannot represent complex relationships (e.g., a property that would concern multiple nodes) or abstraction levels (e.g., an abstract node that represents a subgraph).

In this paper, we propose to extend the property graph model with reification for enhancing its expressivity. Our approach allows to represent complex relationships among entities and to achieve multilevel abstraction in knowledge representation.

The paper is organised as follows. In Sect. 2, we study the different graph models used in knowledge graphs representation. In Sect. 3 we make a study of related works aiming at improving the expressivity of knowledge graphs using the reification process. In Sect. 4, we propose a formal definition of the reified property graphs and we underline the added value of our model. Finally, Sect. 5 concludes the article and presents future work directions.

2 Knowledge Graph Models

A knowledge graph is a flexible way for conceptualizing, representing, and integrating diverse and incomplete real-world knowledge [12,18]. Knowledge graphs can be categorized based on the graph model they use. In this section, we will explore the evolution of knowledge graph models in terms of their expressivity, highlighting limitations in existing models.

A **directed graph** [3] is defined as the triple (V, E, ρ), where V represents a non-empty set of nodes and E constitutes a set of edges connecting nodes in V. $\rho : E \to (V \times V)$ is a total function which gives the couple of nodes associated through an edge.

A **directed edge-labelled graph** (DEL) allows to differentiate multiple edges connecting the same couple of nodes. To do so, it uses labels on edges according to the type of the relationship. A DEL graph is defined as a quadruplet (V, E, ρ, λ), where the additional element $\lambda : E \to \mathcal{L}$ is a total function, with $\mathcal{L}$ being a set of labels.

A **property graph** [2] add properties to nodes and edges and also supports labels for nodes. It is defined as a tuple of 5 components $(V, E, \rho, \lambda, \sigma)$, where $\sigma : (V \cup E) \times Prop \to Val$ is a partial function, where $Prop$ is a finite set of properties and Val is a set of values. If $v \in V$ (resp., $e \in E$), $p \in Prop$, and $\sigma(v, p) = s$ (resp., $\sigma(e, p) = s$), then s is the value of the property p for the node v (resp., the edge e). Compared to the DEL model, $\lambda : (V \cup E) \to 2^{\mathcal{L}}$ assigns a set of labels for each node and edge. In property graphs, aside from labels, both nodes and edges can have properties in the form of property-value pairs. This proves valuable in providing additional metadata and semantics to the nodes and relationships between them [8].

The **hypergraph** model is a generalization of a graph, able to connect more than two nodes. A hypergraph $H = (V, E)$ is defined as a family of hyperedges E, where each hyperedge is a non-empty subset of V, V being finite set of nodes [4]. It has been used in knowledge graphs to represent relationships among multiple entities [9], such as a diploma, a person and a university linked by a graduation.

Until the rise of knowledge graphs, the DEL model was the most used. However, its expressivity is rather limited, as data properties can only be added as a link labelled with the name of the property and a node representing the value of the property. Thus, from a model point of view, the same semantic is used for data properties and for object properties (that link two entities). The property graph model is an improvement regarding the expressivity, as data properties can be directly added as node properties. Nevertheless, there are still plenty of use cases that cannot be easily represented with the property graph model, such as a relationship among more than two entities or a relationship between two links. The hypergraph model overcomes a part of these limits by allowing to link more than two nodes, but a more general concept is needed to greatly improve the expressivity of knowledge graphs. In this context, the process of reification is a good candidate.

3 RDF Reification

Reification is the process of objectifying something and treating it as an element of its own right in the domain of discourse. More specifically, we focus on reification as an abstraction mechanism [13] to enhance the expressivity of graph models. In the literature, we distinguish two categories of reification: Triple-based reification and subgraph-based reification.

The standard RDF reification [14], expressed in the RDF Primer, refers to the process of representing an RDF statement as a new resource, an instance of the `rdf:statement` class, with the main properties `rdf:subject`, `rdf:predicate` and `rdf:object`. In 2014, Nguyen et al. [15] proposed an alternative approach to RDF reification, primarily based on predicates rather than subjects or objects. This approach is the singleton property, which adds a unique predicate to the original predicate of an RDF triple and carries metadata in additional triples related to the unique predicate as a subject. In 2017, Harting [11] proposed the RDF* reification as an extension of standard reification to simplify it. It does not require the creation of a resource to reify a triple, thereby eliminating the multiple statements needed for explicit reification. According to RDF*, the subject or the object of an RDF triple could be itself a triple forming a nested triple, defined by W3C as a *quoted triple*. Beyond the RDF standard, we can cite the work of Rosso *et al.* in 2020 [17]. They proposed to assign to each edge the pair (qualifier, value) representing the name of the metadata and its value. In 2023, Tan *et al.* [20] have represented causal relationships with edges from a Cause node towards an Effect node by adding weight on edges to indicate the strength of the causal effect between nodes. In 2023, Xiong *et al.* [21] proposed nested facts, which involve adding links between links. This approach can be seen as RDF* reification between two triples.

Regarding the second category of reification based on subgraphs, in the scope of RDF, Carroll *et al.* have defined the named graph [6]. A named graph G is a subset of interconnected nodes associated with a unique identifier. It is represented as a pair (G, ID), where ID is a URI (i.e., the identifier of the named graph G). The approach of named graph reification uses the named graph as a context. Based on the idea of named graphs, Stoermer *et al.* [19] proposed to enhance the expressivity of RDF graphs by allowing one or more contexts to be associated with a triple, represented by named graphs. The authors also modelled relationships between contexts with the triple (c_x, R, c_y), where c_x and c_y are the IRIs of two named graphs representing different contexts, and R is the IRI of the relationship between them. Outside of the RDF scope, other approaches have focused on the abstraction of labeled directed graphs. Poulovassilis and Levene [16] proposed the hypernode graph, which represents a subgraph as a node. Since the subgraphs are nodes, they can be connected to each other. Nested graphs [7] is a similar abstraction mechanism.

The studied related works show a need for extending models of knowledge graphs using reification, in order to integrate more advanced knowledge representation techniques. In terms of model expressivity, the standard RDF reification, the singleton property and the RDF* reification are similar. They allow to add information to a triple, the first one by creating a new resource, the second one by directly adding information on the predicate, and the last one by using the triple as is. However, they are not flexible as they require to reify only a triple that means only two nodes. The subgraph based reification technique is less restrictive regarding the selection of elements as it uses a subgraph as input, but it only groups elements and does not include a mechanism to add knowledge about this subgraph as with named graphs, or it does not allow to connect a element of the subgraph with the rest of the graph as with hypernode and nested graphs.

4 A Model for Property Graph Reification

To enhance the expressity of the property graph model, as a model covering most of the capabilities of other graph models, we draw inspiration from the principle of RDF reification to propose a model capable of representing different levels of abstraction. In the following section, we explain the basic strategies of property graph reification, then we propose a formal definition of our model.

4.1 Reification Strategies on Property Graph

As the property graph model allows more advanced representations than an RDF graph, these evolutions have to be considered before applying reification techniques on property graphs. First, while data properties and object properties are both represented with an edge in RDF, within a property graph, a data property becomes a node property. Thus, reification on the property graph model must include node properties to keep at least the same level of expressivity. Second, in a property graph, each node and edge can have multiple labels, rather than just one in a RDF graph. Therefore, the reification mechanism must consider the labels for nodes and edges.

4.2 Formal Model for Property Graph Reification

To define reification on property graphs, we start from the formal definition of the property graph model and extend it. Therefore, a property graph model G with reification is defined by a tuple

$$G = (V, E, R, \rho, \lambda, \sigma, \alpha)$$

where:

- $R \subseteq V$, R is a subset of V containing the reified nodes;
- $\alpha : R \rightarrow SG$ is a total function mapping the reified nodes to their corresponding subgraph. SG is a finite set of tuples $sg = (V_r, E_r, R_r, \rho_r, \lambda_r, \sigma_r, \alpha_r)$. $r \in R$ and $sg \in SG$, then $\alpha(r) = sg$ where sg is the subgraph of the reified node r.
  In the definition of sg, $V_r \subseteq V$, $E_r \subseteq E$, $R_r \subseteq R$, $\rho_r = \rho|_{E_r}$, $\lambda_r = \lambda|_{V_r \cup E_r}$ and $\lambda_r : (V_r \cup E_r) \rightarrow 2^{\mathcal{L}_r}$, $\sigma_r = \sigma|_{(V_r \cup E_r) \times Prop_r}$, $\alpha_r = \alpha|_{R_r}$. The functions $\rho_r, \lambda_r, \sigma_r$ and α_r are restricted functions. $Prop_r$ is a finite set of selected properties of the subgraph and $\mathcal{L}_r$ is a finite set of selected labels, thus $Prop_r \subseteq Prop$ and $\mathcal{L}_r \subseteq \mathcal{L}$.

Our reification approach applied to property graphs provides a powerful mechanism for representing complex relationships and multiple abstraction levels within the graph. By selecting a subset of nodes along with their labels, properties and edges, we can effectively reify them and represent them with a new node corresponding to a subgraph $sg \in SG$. This mechanism is recursive: a reified node can also represent other reified nodes.

A reified node $r \in R$ benefits from all the modelling capabilities of standard nodes as $R \subseteq V$. It allows to define labels and properties for reified nodes, as well as edges whether the destination node is a standard or a reified node.

As α gives a subgraph for each reified node that fits the definition of our property graph extended model, it enables recursive reifications in which a reified node could itself contain other reified nodes. From a representation perspective, it means that the definition of our model allows to define several levels of abstraction, and to navigate among them, thus enhancing the expressivity and modelling capabilities of knowledge graphs.

4.3 Reified Node Specification

To build a reified node, users must specify which elements are concerned by the reification. As stated by the model definition, the model allows to reify on nodes, links, labels and properties. Thus, the building function must accept these elements individually or any combination of them.

To do so, we propose the following function:

$$\beta(V_r, E_r, \lambda_r, \sigma_r) = r$$

where V_r is the set of selected nodes, E_r the set of selected edges, $\lambda_r : (V_r \cup E_r) \rightarrow 2^{\mathcal{L}_r}$ restricts the labels kept, and $\sigma_r : (V_r \cup E_r) \times Prop_r \rightarrow Val$ restricts the properties kept.

As λ_r and σ_r have their domain restricted on the selected nodes and edges, it guarantees that the selected labels and properties belong to an existing node or edge of the reification. Furthermore, it is a flexible mechanism as it allows to keep a given property or label for a specific node or edge but not necessarily for all the nodes and edges that have this property or label. $\rho_r = \rho|_{E_r}$ is built automatically from E_r, by restricting the function ρ on E_r.

To obtain R_r stated in the model, we rely on R of the global graph, as reified nodes of a subgraph are also reified nodes in the graph containing the subgraph. Therefore, $R_r = \{v | v \in R \cap V_r\}$.

5 Conclusion and Perspectives

Representing real-world data using knowledge graphs often requires complex relationships. The property graph model stands out as a particularly concise approach compared to other graph models for representing knowledge. Its strength lies in its ability to represent knowledge with a reduced structural size using labels and properties on nodes and edges. However, this model does have its limitations, notably its inability to represent complex relationships between nodes and its inability to define multiple levels of abstraction. To overcome these limitations, we extend the current property graph model and propose a new model supporting reification. We have formalised our reification approach that empowers the property graph model to model complex relationships by representing a sub-graph of a set of nodes, edges as well as their related properties and labels in the form of a node. This node can itself have properties, labels and edges. Furthermore, this mechanism is recursive and enables multilevel knowledge abstraction.

Our forthcoming work is structured around two main axes. Firstly, we plan to proceed with the technical implementation of our approach. Secondly, we intend to adapt current property graph query languages to our model. This adaptation is crucial to enable effective management of multilevel abstraction when querying knowledge graphs. For example, queries will need to be capable of aggregating properties, particularly when calculating a property based on the properties of the constituent elements of an abstract node. These two axes represent a crucial step in fully exploiting the potential of our model in practical contexts and facilitating its integration into existing applications and information systems.

References

1. Abiteboul, S., et al.: Research directions for principles of data management. Dagstuhl Manifestos **7**(1), 1–29 (2018)
2. Angles, R., Arenas, M., Barceló, P., Hogan, A., Reutter, J., Vrgoč, D.: Foundations of modern query languages for graph databases. ACM Comput. Surv. (CSUR) **50**(5), 1–40 (2017)
3. Bang-Jensen, J., Gutin, G.: Classes of Directed Graphs, vol. 11. Springer, Cham (2018). https://doi.org/10.1007/978-3-319-71840-8
4. Berge, C.: Graphs and Hypergraphs. North-Holland, Amsterdam (1973)
5. Bernstein, P.A., Halevy, A.Y., Pottinger, R.A.: A vision for management of complex models. ACM SIGMOD Rec. **29**(4), 55–63 (2000)
6. Carroll, J.J., Bizer, C., Hayes, P., Stickler, P.: Named graphs. J. Web Semant. **3**(4), 247–267 (2005)
7. Chein, M., Mugnier, M.L., Simonet, G.: Nested graphs: a graph-based knowledge representation model with FOL semantics. In: KR, pp. 524–535 (1998)
8. Das, S., Srinivasan, J., Perry, M., Chong, E.I., Banerjee, J.: A tale of two graphs: property graphs as RDF in oracle. In: International Conference on Extending Database Technology (EDBT), pp. 762–773 (2014)
9. Fatemi, B., Taslakian, P., Vazquez, D., Poole, D.: Knowledge hypergraphs: prediction beyond binary relations. In: Bessiere, C. (ed.) Proceedings of the Twenty-Ninth International Joint Conference on Artificial Intelligence, IJCAI 2020, pp. 2191–2197. International Joint Conferences on Artificial Intelligence Organization, July 2020. https://doi.org/10.24963/ijcai.2020/303, main track
10. Gutiérrez, C., Sequeda, J.F.: Knowledge graphs. Commun. ACM **64**(3), 96–104 (2021)
11. Hartig, O.: Foundations of RDF* and SPARQL*: (an alternative approach to statement-level metadata in RDF). In: AMW 2017 11th Alberto Mendelzon International Workshop on Foundations of Data Management and the Web, Montevideo, Uruguay, 7–9 June 2017, vol. 1912. Juan Reutter, Divesh Srivastava (2017)
12. Hogan, A., et al.: Knowledge graphs. ACM Comput. Surv. (CSUR) **54**(4), 1–37 (2021)
13. Malenfant, J., Jacques, M., Demers, F.: A tutorial on behavioral reflection and its implementation. In: Proceedings of the Reflection, vol. 96, pp. 1–20 (1996)
14. Manola, F., Miller, E.: RDF reification (2004). https://www.w3.org/TR/rdf-primer/#reification
15. Nguyen, V., Bodenreider, O., Sheth, A.: Don't like RDF reification? Making statements about statements using singleton property. In: Proceedings of the 23rd International Conference on World Wide Web, pp. 759–770 (2014)
16. Poulovassilis, A., Levene, M.: A nested-graph model for the representation and manipulation of complex objects. ACM Trans. Inf. Syst. (TOIS) **12**(1), 35–68 (1994)
17. Rosso, P., Yang, D., Cudré-Mauroux, P.: Beyond triplets: hyper-relational knowledge graph embedding for link prediction. In: Proceedings of the Web Conference 2020, pp. 1885–1896 (2020)
18. Sequeda, J., Lassila, O.: Building enterprise knowledge graphs. In: Designing and Building Enterprise Knowledge Graphs, pp. 97–128. Springer, Cham (2021). https://doi.org/10.1007/978-3-031-01916-6_4
19. Stoermer, H., et al.: RDF and contexts: use of SPARQL and named graphs to achieve contextualization. In: Proceedings of 2006 Jena User Conference (2006)

20. Tan, F.A., Paul, D., Yamaura, S., Koji, M., Ng, S.K.: Constructing and interpreting causal knowledge graphs from news. arXiv preprint arXiv:2305.09359 (2023)
21. Xiong, B., Nayyeri, M., Luo, L., Wang, Z., Pan, S., Staab, S.: NestE: modeling nested relational structures for knowledge graph reasoning. arXiv preprint arXiv:2312.09219 (2023)
22. Zou, X.: A survey on application of knowledge graph. J. Phys. Conf. Ser. **1487**, 012016 (2020)

ExpO: Towards Explaining Ontology-Driven Conceptual Models

Elena Romanenko[1(✉)], Diego Calvanese[1,2], and Giancarlo Guizzardi[3,4]

[1] Free University of Bozen-Bolzano, Bolzano, Italy
{eromanenko,diego.calvanese}@unibz.it
[2] Umeå University, Umeå, Sweden
[3] University of Twente, Enschede, The Netherlands
g.guizzardi@utwente.nl
[4] Stockholm University, Stockholm, Sweden

Abstract. Ontology-driven conceptual models play an explanatory role in complex and critical domains. However, since those models may consist of a large number of elements, including concepts, relations and sub-diagrams, their reuse or adaptation requires significant efforts. While conceptual model engineers tend to be biased against the removal of information from the models, general users struggle to fully understand them. The paper describes ExpO—a prototype that addresses this trade-off by providing three components: *(1)* an API that implements model transformations, *(2)* a software plugin aimed at modelers working with the language OntoUML, and *(3)* a web application for model exploration mostly designed for domain experts. We describe characteristics of every component and specify scenarios of possible usages.

Keywords: Ontology-Driven Conceptual Models · OntoUML · Pragmatic explanation · Software Tools

1 Introduction

A *conceptual model* (CM) is an abstract, high-level representation of the domain of interest, the task that needs to be carried out, or the software itself. In recent years, *ontology-driven conceptual models* (ODCMs) have been proposed as a particular class of conceptual models that gain an advantage by utilizing ontological theories to develop engineering artefacts [15].

Both ODCMs and traditional CMs alike play a fundamental role in organizing communication between people with different backgrounds, such as programmers, ontology engineers, and domain experts. Models are also supposed to be easily reused and extended. Unfortunately, in reality, there is an unspoken disagreement between domain experts and conceptual model engineers—authors of the existing ODCMs. While modelers tend to consider all information specified in the model as necessary, thus, resisting attempts to simplify these artefacts (see experiments in [12]), domain experts, as general users, are struggling to fully understand how to adapt an already existing model to their needs.

J. Araújo et al. (Eds.): RCIS 2024, LNBIP 514, pp. 20–28, 2024.
https://doi.org/10.1007/978-3-031-59468-7_3

In this paper, we present *ExpO* — a prototype system for explaining existing ODCMs. We rely here on the notion of *pragmatic explanation* for ODCMs as discussed in [11] for domain ontologies, and focus on explaining models without changing the form of explanation. ExpO consists of *(1)* an API that implements model transformations that are supposed to help in ODCM understanding, *(2)* an extension of the existing OntoUML Plugin with the API functionality, and *(3)* a web application for model exploration mostly designed for domain experts with a non-technical background. This work is based on several previous papers [5, 7,10,11]. However, our main purpose here is not to formulate a final set of explanation operations needed to reach a better understanding of a given model, but to present the system itself.

The remainder of the paper is organized as follows: Sect. 2 presents background; Sect. 3 proposes ExpO, introduces its architecture, and elaborates on various components of the system; Sect. 4 provides final considerations and outlines future work.

2 Background

Foundational ontologies are a special class of ontologies specifying a general schema for describing objects, their constitution and composition, roles that objects can play, events and their goals, and qualities of objects and events. In a recent special issue of the Applied Ontology journal [1], seven of the most commonly used foundational ontologies were listed, including the *Unified Foundational Ontology* (UFO).

UFO [4] draws contributions from Formal Ontology in Philosophy, Philosophical Logic, Cognitive Psychology, and Linguistics. It is particularly interesting for our research given that it is used as a foundation for OntoUML, one of the most widely used languages in ontology-driven conceptual modeling [15]. *OntoUML* is a language that extends UML class diagrams by defining a set of stereotypes. These stereotypes expand UML's meta-model so that classes and associations decorated with them bring precise (real-world) semantics grounded in UFO. Additionally, OntoUML models satisfy a number of semantically motivated syntactic constraints, ensuring their compliance with the UFO axiomatization [6].

In general, conceptual model engineers are better supported for model development than domain experts for model exploration. For example, *Visual Paradigm* (VP)[1] is a widespread modeling tool that facilitates the development of different types of diagrams, including UML models. For those who are developing ODCMs, there is, e.g., an *OntoUML Plugin* for VP[2] that automatically checks the ontological consistency of an OntoUML diagram in light of UFO. Nevertheless, both proficiency in technical skills and familiarity with UFO, along with the obligation to install specialized software, are essential for utilizing the plugin. At the same time, there is no proper tool for users who would like to

[1] https://www.visual-paradigm.com.

[2] https://github.com/OntoUML/ontouml-vp-plugin.

familiarize themselves with the existing model or simply check if it is suitable for their specific requirements.

To facilitate understanding of existing models, software tools that manage ODCMs should incorporate features that provide explanations for such models. In this context, the term 'explanation' refers to *pragmatic explanation*, where 'pragmatic' signifies an 'instrumentalist' approach to constructing explanations (see [16]), emphasizing the creation of a 'toolbox' designed to assist users in reaching their goals. In [11], we show that complexity management approaches, such as clustering [7] and abstraction [5,10], can be viewed as pragmatic explanation techniques when talking about domain ontologies. Recent experiments revealed that this is also true for ODCMs [12]. However, to the best of our knowledge, there is no uniform tool that offers different transformations of ODCMs, potentially enhancing comprehension of the given model.

Visual Paradigm is a popular modeling tool, but together with the OntoUML Plugin, it is more used by conceptual modelers and ontology engineers for model development rather than by domain experts for model exploration. Currently, it already provides some desired functionality, such as model clustering [7].

Protégé[3] [9] aims at OWL artefacts. ODCMs—exported into Turtle format—may be examined in it with different visualization plugins, e.g., *OntoGraph*[4]. Although these plugins lack important functionality, they provide the possibility to focus or hide the concept of interest. This, however, may lead to a cognitive load of users when dealing with large models [8].

WebVOWL[5] also can load a model in Turtle format, but visualizes individuals as lists, which leads to the same problem of cognitive overload.

Although *Evonne*[6] [8] was developed with a completely different goal—to support interactive debugging of ontologies—it seems to be the most relevant tool to our research since it is aimed at explanations. At the same time, Evonne is aimed at OWL models and, hence, does not support explanation and complexity management for ODCMs in a manner that is aligned with foundational categories.

3 ExpO Architecture

Before presenting the design of the ExpO system, we first need to determine its potential users and their requirements. During a user study we carried out [12], it became obvious that at least two categories of users must be distinguished: *(1) authors of the models* (model engineers, experienced modelers), and *(2) domain experts* without prior knowledge about OntoUML modeling or potential model users who are unfamiliar with the domain.

Taking the defined categories of users into account, we formulated the following design goals for the system.

[3] http://protege.stanford.edu.
[4] https://protegewiki.stanford.edu/wiki/OntoGraf.
[5] http://vowl.visualdataweb.org/webvowl.html.
[6] https://imld.de/en/research/research-projects/evonne.

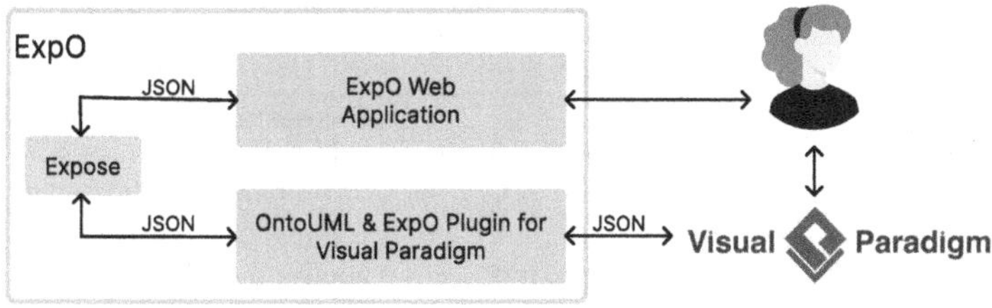

Fig. 1. Broad overview of the approach.

DG1: *Support different levels of expertise.* For users with less modeling experience the system should remain intuitive.
DG2: *Minimize setup difficulties.* Novice users with little technical background should receive a ready-to-use tool.
DG3: *Enable interactive exploration.* Interaction plays a critical role in providing explanations and reaching understanding [8, 11].
DG4: *Build on familiar representations.* The system should use standard visual representations of graphs as node-link diagrams.
DG5: *Keep the original layout.* The system should try to keep the original (expert) layout of the ODCM when possible in order not to confuse users, who probably worked with the same model in VP before.

In [14], the author suggests the following strategy for visual information-seeking: "overview first, zoom and filter, then details on demand". Based on this guideline, seven tasks were suggested for systems that provide visual content:

Overview: gain an overview of the entire collection of elements;
Zoom: zoom in on items of interest;
Filter: filter out uninteresting items;
Details-on-demand: select an item or group and get details when needed;
Relate: view relations among items;
History: keep a history of actions to support undo and replay;
Extract: allow extraction of sub-collections.

This approach was further extensively reused, e.g., in [2]. During the development of our system, we took into account these guidelines and the refined associated tasks, and suggested an architecture of the system with three separate components (see Fig. 1):

- *Expose*, a server that provides an API for ODCM transformations;
- *OntoUML & ExpO Plugin for VP*, an updated version of the OntoUML Plugin for VP with extended functionality;
- *ExpO Web Application*, a web interface mostly aimed at domain experts.

The Expose component is written in Python using the FastAPI. It is a server component[7], that provides open routes that implement transformation opera-

[7] Use https://w3id.org/ExpO/expose/health to check if the server is accepting requests.

Table 1. Short description of some of the API routes.

Route	Specification		Description
Focus	*Type* *Parameters*	POST Body: graph, node, hop	Focuses on the given node and keeps only those concepts that are reached by no more than hop relations.
Cluster	*Type* *Parameters*	POST Body: graph, node	If the given node is a *Relator*, applies the relator-centric approach for clustering [7].
Define	*Type* *Parameters*	GET Parameters: concept, number of definitions	Provides several definitions of the given concept that can be found in a dictionary. This route is used by the web application only.
Expand	*Type* *Parameters*	POST Body: graph, node, limit	Finds a similar concept (by name and stereotype) in the OntoUML / UFO Catalog [13] and, if found, extends the hierarchy with the data from the Catalog using no more than limit concepts.
Abstract	*Type* *Parameters*	POST Body: graph, abstraction type	Applies the given type of abstraction (one or more) to the graph. The supported abstraction types are "parthood", "hierarchy", and "aspects" [10].
Fold	*Type* *Parameters*	POST Body: graph, node	Collapses all hierarchical and part-whole relations of the given node.

tions[8]. A short description of some of the open routes as well as additional remarks regarding some of the requests are collected in Table 1.

The *Expand* route keeps an index of all concepts and their stereotypes that were mentioned in all models of the OntoUML / UFO Catalog [13]. In order to get the models with the concept of interest, it sends an API request to the GitHub REST API[9] and tries to extend the current model with the collected information. The *Define* route sends an API request to the *Wiktionary*[10] service to collect likely definitions of the given concept. We assume that this information can be used by novice users not familiar with the domain. All other routes are processed directly by Expose, and, as shown in Table 1, most of them are POST requests. The reason for this is that Expose does not store user data. Thus, the general idea is to take the original model, apply the requested transformation operation to it, and return the result in the right format. All routes respond with JSON, the content of which depends on the type of the output format. For the plugin, the server returns only the model that can be loaded directly into

[8] The full documentation can be found in the corresponding folder of the project on https://w3id.org/ExpO/github and on https://w3id.org/ExpO/expose/docs.

[9] https://docs.github.com/en/rest?apiVersion=2022-11-28.

[10] https://en.wiktionary.org.

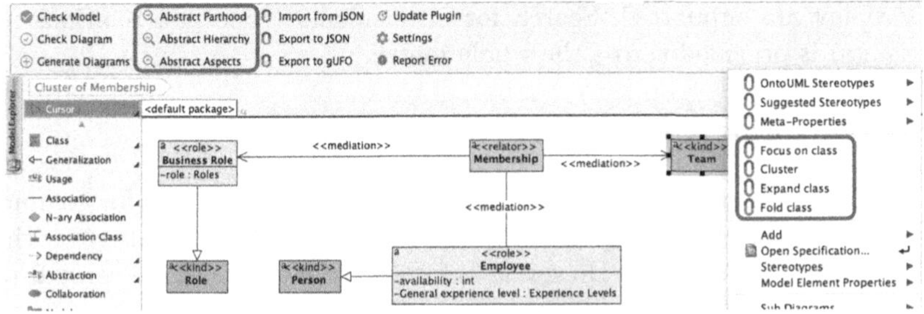

Fig. 2. Changes in the main menu and the context menu of the plugin. In the picture, you can see the result of applying a *Cluster* request to the 'Membership' concept for the model 'jacobs2022sdpontology' from the OntUML / UFO Catalog [13].

VP. For the web application, the result consists of the original graph (for future processing) and its simplification and adaptation for the web interface.

The original OntoUML Plugin for VP was updated in order to support the functionality provided by Expose. Figure 2 shows the updates in the main and context menus that were introduced. The plugin is developed in Java and published on GitHub[11]. However, the user interface is constrained by VP. For instance, while it supports zoom functionality for models, there are limitations when searching for a concept or relation. Users can only see those elements that were found on the active diagram, without the ability to search across the entire project. Additionally, the system has some difficulties in maintaining a comprehensive history of actions; although actions are mostly reversible within a session, each model received from Expose is treated as an entirely new one. These limitations are intrinsic to the existing implementations of VP and the original plugin and cannot be readily overcome. For these reasons, achieving the first three design goals (DG1–DG3) is challenging using only the plugin. For users who do not want to install additional software, miss the required technical skills, or just want to quickly explore the model without the necessity to download it, we suggest using the ExpO Web Application.

The ExpO Web Application[12] is written in JavaScript with the help of the React library[13] and the react-d3-graph component[14]. In order not to confuse the user (who could have used the OntoUML & ExpO Plugin before) and in accordance with the last design goal, the smart colouring scheme (see, e.g., in [7]) and the layout are kept as close to the original model as possible. Note, that full compliance cannot always be achieved, since VP provides an opportunity to divide the project into several models, while the web application combines all of them in one view. Both zoom opportunities and the possibility to perform

---

[11] https://w3id.org/ExpO/plugin.
[12] https://w3id.org/ExpO.
[13] https://react.dev.
[14] https://danielcaldas.github.io/react-d3-graph.

undo / replay are supported. Search for all elements (concepts, relations, and constraints) is provided across the whole model.

4 Conclusions and Future Work

Ontology-driven conceptual models are supposed to build a bridge for communication between ontology engineers or programmers and domain experts without special technical knowledge. However, before these models can be reused, they need to be understood by their users.

In this paper, we have presented ExpO, a prototype system aimed at producing explanations for ODCMs. The ExpO system consists of three components, each with its own functionality. The *ExpO Server* applies transformation operations to ODCMs. Those operations cover most of the tasks described by the "Visual Information-Seeking Mantra". The rest of the tasks, namely Zoom and History, are covered by the *OntoUML & ExpO Plugin for VP* and by the *ExpO Web Application*. The former is mostly aimed at professionals, who have worked with modeling tools before, while the web application does not require installation or model download, and allows for model investigation on-the-fly.

We expect to continue to extend the functionality of the system. A prospective method can be based on showing the hierarchy below / above the selected concept. For now, most of the explanation transformations are applied in an automatic mode, but the user may be interested in selecting certain model elements that need to be maintained in the final explanation. Also, ideally, each ODCM should be accompanied by a list of *competency questions* [3]. By filtering this list according to concepts that are still left in view, we can help our users determine how far they would like to reduce the model. Finally, a qualitative user study is needed, investigating whether this set of transformation operations is enough to support a full understanding of the model, and the role the web application plays in it.

Resource Availability Statement: The ExpO Server is available at https://w3id.org/ExpO/expose/health. The ExpO Plugin can be downloaded and installed from the GitHub repository https://w3id.org/ExpO/plugin. The ExpO Web Application is available at https://w3id.org/ExpO. The corresponding GitHub repository for the project is https://w3id.org/ExpO/github. The software is distributed under the Apache 2.0 license.

Acknowledgements. This research has been partially supported by the Province of Bolzano and DFG through the project D2G2 (DFG grant n. 500249124), by the HEU project CyclOps (grant agreement n. 101135513), and by the Wallenberg AI, Autonomous Systems and Software Program (WASP), funded by the Knut and Alice Wallenberg Foundation.

References

1. Borgo, S., Galton, A., Kutz, O.: Foundational ontologies in action. Appl. Ontol. **17**, 1–16 (2022). https://doi.org/10.3233/AO-220265

2. Golfarelli, M., Pirini, T., Rizzi, S.: Goal-based selection of visual representations for big data analytics. In: de Cesare, S., Frank, U. (eds.) ER 2017. LNCS, vol. 10651, pp. 47–57. Springer, Cham (2017). https://doi.org/10.1007/978-3-319-70625-2_5
3. Grüninger, M., Fox, M.S.: Methodology for the design and evaluation of ontologies. In: Proceedings of the IJCAI 1995 Workshop on Basic Ontological Issues in Knowledge Sharing (1995). http://www.eil.utoronto.ca/wp-content/uploads/enterprise-modelling/papers/gruninger-ijcai95.pdf
4. Guizzardi, G., Botti Benevides, A., Fonseca, C.M., Porello, D., Almeida, J.P.A., Sales, T.P.: UFO: unified foundational ontology. Appl. Ontol. **17**(1), 167–210 (2022). https://doi.org/10.3233/AO-210256
5. Guizzardi, G., Figueiredo, G., Hedblom, M.M., Poels, G.: Ontology-based model abstraction. In: Proceedings of the 13th International Conference on Research Challenges in Information Science (RCIS), pp. 1–13. IEEE (2019). https://doi.org/10.1109/RCIS.2019.8876971
6. Guizzardi, G., Fonseca, C.M., Almeida, J.P.A., Sales, T.P., et al.: Types and taxonomic structures in conceptual modeling: a novel ontological theory and engineering support. Data Knowl. Eng. **134**, 101891 (2021). https://doi.org/10.1016/j.datak.2021.101891
7. Guizzardi, G., Sales, T.P., Almeida, J.P.A., Poels, G.: Automated conceptual model clustering: a relator-centric approach. Softw. Syst. Model. **21**, 1363–1387 (2022). https://doi.org/10.1007/s10270-021-00919-5
8. Méndez, J., Alrabbaa, C., Koopmann, P., Langner, R., et al.: Evonne: a visual tool for explaining reasoning with OWL ontologies and supporting interactive debugging. Comput. Graph. Forum (2023). https://doi.org/10.1111/cgf.14730
9. Musen, M.A.: The Protégé project: a look back and a look forward. AI Matters **1**(4), 4–12 (2015). https://doi.org/10.1145/2757001.2757003
10. Romanenko, E., Calvanese, D., Guizzardi, G.: Abstracting ontology-driven conceptual models: Objects, aspects, events, and their parts. In: Guizzardi, R., Ralyté, J., Franch, X. (eds.) RCIS 2022. LNBIP, vol. 446, pp. 372–388. Springer, Cham (2022). https://doi.org/10.1007/978-3-031-05760-1_22
11. Romanenko, E., Calvanese, D., Guizzardi, G.: Towards pragmatic explanations for domain ontologies. In: Corcho, O., Hollink, L., Kutz, O., Troquard, N., Ekaputra, F.J. (eds) EKAW 2022. LNAI, vol. 13514, pp. 201–208. Springer, Cham (2022). https://doi.org/10.1007/978-3-031-17105-5_15
12. Romanenko, E., Calvanese, D., Guizzardi, G.: What do users think about abstractions of ontology-driven conceptual models? In: Nurcan, S., Opdahl, A.L., Mouratidis, H., Tsohou, A. (eds) RCIS 2023. LNBIP, vol. 476, pp. 53–68. Springer, Cham (2023). https://doi.org/10.1007/978-3-031-33080-3_4
13. Sales, T.P., Barcelos, P.P.F., Fonseca, C.M., Valle Souza, I., et al.: A FAIR catalog of ontology-driven conceptual models. Data Knowl. Eng. **147**, 102210 (2023). https://doi.org/10.1016/j.datak.2023.102210
14. Shneiderman, B.: The eyes have it: a task by data type taxonomy for information visualizations. In: Proceedings of the 1996 IEEE Symposium on Visual Languages, pp. 336–343. IEEE Computer Society (1996)

15. Verdonck, M., Gailly, F.: Insights on the use and application of ontology and conceptual modeling languages in ontology-driven conceptual modeling. In: Comyn-Wattiau, I., Tanaka, K., Song, I.-Y., Yamamoto, S., Saeki, M. (eds.) ER 2016. LNCS, vol. 9974, pp. 83–97. Springer, Cham (2016). https://doi.org/10.1007/978-3-319-46397-1_7
16. Weber, E., Van Bouwel, J., De Vreese, L.: Scientific Explanation. Springer Briefs in Philosophy. Springer, Dordrecht (2013). https://doi.org/10.1007/978-94-007-6446-0

Translucent Precision: Exploiting Enabling Information to Evaluate the Quality of Process Models

Harry Herbert Beyel(✉) and Wil M. P. van der Aalst

Chair of Process and Data Science, RWTH Aachen University, Aachen, Germany
{beyel,wvdaalst}@pads.rwth-aachen.de

Abstract. An event log stores information about executed activities in a process. Conformance-checking techniques are used to measure the quality of a process model using an event log. Part of the investigated quality dimensions is *precision*. Precision puts the behavior of a log and a model in relation. There are event logs that also store information about *enabled* activities besides the actual executed activities. These event logs are called *translucent event logs*. A technique for measuring precision is escaping arcs. However, this technique does not consider information on enabled activities contained in a translucent event log. This paper provides a formal definition of how to compute a precision score by considering translucent information. We discuss our method using a translucent event log and four different models. Our translucent precision score conveys the underlying concept by considering more information.

Keywords: Process Mining · Conformance Checking · Precision

1 Introduction

In each organization, processes play a vital role. The execution of processes may leave event data in information systems. Typically, an event consists of three attributes: a *case identifier*, an *activity*, and a *timestamp*. We call a collection of these data an *event log*. Such event logs are used in *process mining* [1]. Conformance checking, an area of process mining, consists of various quality dimensions, including *precision* [12]. Precision evaluates whether a process model allows for more behavior than captured in the event log. Suppose L describes the behavior in an event log, and M captures the behavior contained in a process model. In that case, we can define the general idea of precision as follows:[1]

$$precision = \frac{|L \cap M|}{|M|}$$

We thank the Alexander von Humboldt (AvH) Stiftung for supporting our research.

[1] The notion of "behavior" is left vague here. There is the challenge that the event log is a finite sample, but the model may describe infinitely many traces (due to loops).

J. Araújo et al. (Eds.): RCIS 2024, LNBIP 514, pp. 29–37, 2024.
https://doi.org/10.1007/978-3-031-59468-7_4

Table 1. Example translucent event log.

Event	Case	Activity	Enabled Activities	Time-stamp
e_1	1	a	{a}	13:37:37
e_2	1	b	{b, c}	13:37:38
e_3	1	c	{c}	13:37:39
e_4	1	e	{d, e}	13:37:40
e_5	2	a	{a}	13:37:41
e_6	2	c	{b, c}	13:37:42
e_7	2	b	{b}	13:37:43
e_8	2	d	{d, e}	13:37:44

Event	Case	Activity	Enabled Activities	Time-stamp
e_9	2	b	{b, c}	13:37:45
e_{10}	2	c	{c}	13:37:46
e_{11}	2	e	{d, e}	13:37:47
e_{12}	3	a	{a}	13:37:48
e_{13}	3	c	{b, c}	13:37:49
e_{14}	3	b	{b}	13:37:50
e_{15}	3	e	{d, e}	13:37:51

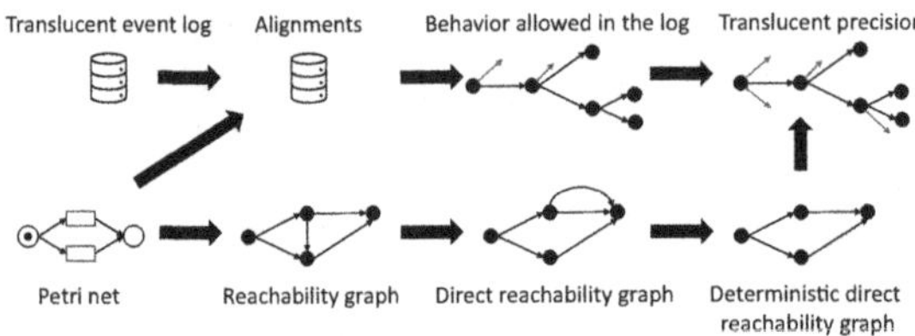

Fig. 1. Sketch of our approach to measure translucent precision.

Besides capturing only executed activities, an event can capture information on *enabled activities*. If an event log consists of such events, we call the event log a *translucent event log* [2]. A translucent event log can be, e.g., created when tasks performed in a desktop environment are captured to create training data for software bots [9]. An example of a translucent event log is shown in Table 1. When measuring precision between a translucent event log and a process model, it is vital to consider the information on enabled activities. In this paper, we present the first approach to a precision method that considers the information captured in translucent event logs and fits the intuitive meaning of precision. Our approach is based on escaping arcs [19]. An overview is depicted in Fig. 1.

2 Related Work

A technique for measuring precision is based on escaping arcs [3,7,8,19]. Given an event log, a prefix automaton is built. Traces are replayed on the model, and it is checked whether the model allows for more behavior than in the automaton. Another approach is based on anti-alignments [15]. The previous approach does not capture model deviation if it is not directly involved in the replay. This approach aims to solve this issue. An anti-alignment is an execution sequence in a given model that significantly differs from all traces in the log [13,14]. Another approach relies on negative events [16]. Negative events are sets of events that were prohibited from taking place. Such events are induced for each position in the event log. [20] introduces behavioral precision. [10,11] refine the approach. There exist stochastic-aware precision measures [18] and approaches for object-centric process mining [5]. However, none of the presented approaches uses information on enabled activities provided in the event log.

3 Preliminaries

Definition 1 (Sets, Powersets, Multisets, Sequences). *Given a set X and a function f, $f(X) = \{f(x) \mid x \in X\}$ denotes applying the function f on all elements of set X. For sets X and Y, $X \times Y = \{(x, y) \mid x \in X, y \in Y\}$ denotes the cartesian product. The powerset of a set X is denoted as $\mathcal{P}(X) = \{X' \mid X' \subseteq X\}$. $\mathcal{B}(X)$ denotes the set of all multisets over set X. E.g., if $X = \{x, y, z\}$, a possible bag is $[x, x, y] = [x^2, y]$. Given a set X, a sequence $\sigma = \langle \sigma_1, \ldots, \sigma_n \rangle$, $\sigma \in X^*$, denotes a sequence over X. σ_i denotes the sequence's i-th element. The length of a sequence σ is denoted as $|\sigma|$. Given a sequence $\sigma = \langle \sigma_1, \ldots, \sigma_{|\sigma|} \rangle$ and a function f, $f(\sigma) = \langle f(\sigma_1), \ldots, f(\sigma_{|\sigma|}) \rangle$. $pref_i(\sigma) = \langle \sigma_1, \ldots, \sigma_i \rangle$ refers to the prefix of a sequence containing the first i elements. $pref_0 = \langle \rangle$. Given a sequence σ and a set X', $\sigma\restriction_{X'}$ denotes a sequence projections, e.g., $\langle a, b, c, d \rangle\restriction_{\{a,c\}} = \langle a, c \rangle$.*

Translucent event logs capture information on enabled activities in addition to executed ones. Hence, the executed activity must also be enabled in the corresponding event. Also, we assume that all enabled activities in an event log are performed at some point. $\mathcal{U}_{case}$ is the universe of case identifiers, $\mathcal{U}_{act}$ is the universe of activity names, and $\mathcal{U}_{time}$ is the universe of timestamps.

Definition 2 (Translucent Event Log, Trace). *$\mathcal{U}_{ev}$ is the universe of events. $e \in \mathcal{U}_{ev}$ is an event, $\pi_{case}(e) \in \mathcal{U}_{case}$ is the case of e, $\pi_{time}(e) \in \mathcal{U}_{time}$ is the time of e, $\pi_{en}(e) \subseteq \mathcal{U}_{act}$ are the enabled activities of e, $\pi_{act}(e) \in \pi_{en}(e)$ is the activity of e. In addition, $\bigcup_{e \in L} \pi_{en}(e) = \bigcup_{e \in L} \{\pi_{act}(e)\}$. A translucent event log L is a set of events $L \subseteq \mathcal{U}_{ev}$. For simplicity, we assume that events in L are totally ordered s.t. for $e_1, e_2 \in L$, $e_1 < e_2$ implies $\pi_{time}(e_1) \leq \pi_{time}(e_2)$. A trace is a sequence of all events of a case ordered from earliest to latest, i.e., $\sigma^{L,c} = \langle e_1, \ldots, e_n \rangle$, s.t. for $c \in \pi_{case}(L)$, $\{e_1, \ldots, e_n\} = \{e \in L \mid \pi_{case}(e) = c\}$ and $e_1 < \ldots < e_n$. The set of traces of L is denoted as $\Sigma^L = \{\sigma^{L,c} \mid c \in \pi_{case}(L)\}$.*

For the example translucent event log shown in Table 1, $\pi_{case}(e_2) = 1$, $\pi_{act}(e_2) = b$, $\pi_{en}(e_2) = \{b, c\}$, and $\pi_{time}(e_2) =$ 13:37:38, and $\Sigma^L = \{\langle e_1, e_2, e_3, e_4 \rangle, \ldots\}$.

Definition 3 (Marked Labeled Petri Net). *A labeled Petri net is a tuple $N = (P, T, F, A, l)$, where P is a set of places, T is a set of transitions s.t. $P \cap T = \emptyset$, and $F \subseteq (T \times P) \cup (P \times T)$ is a set of directed arcs. $A \subseteq \mathcal{U}_{act} \cup \{\tau\}$ is a set of activity labels, and $l : T \to A$ is a labeling function where τ denotes the activity of silent transitions. A marking $M \in \mathcal{B}(P)$ is a multiset of places. We write (N, M) to refer to the Petri net N in marking M.*

We focus on sound workflow nets [4]. Petri net firing rules can be found in [1].

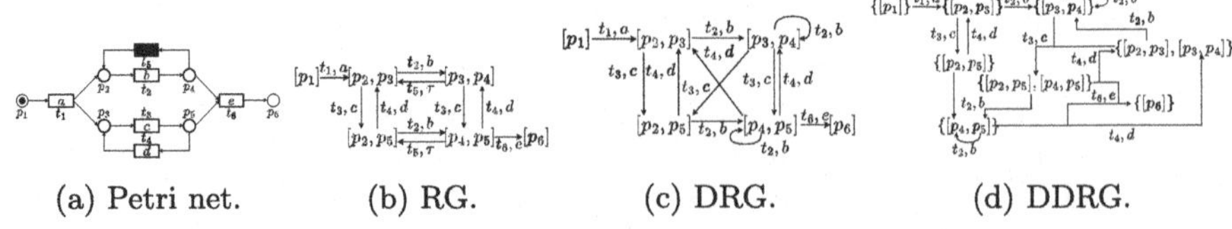

(a) Petri net. (b) RG. (c) DRG. (d) DDRG.

Fig. 2. Structures based on the event log shown in Table 1.

Definition 4 (Firing Sequence). *Let $N = (P, T, F, A, l)$ be a Petri net. The successive firing of all transitions in $\sigma \in T^*$ is denoted as $(N, M_1) \xrightarrow{\sigma} (N, M_{n+1})$. Let $M_{in} \in \mathcal{B}(P)$ be the initial marking. We define a marking $M' \in \mathcal{B}(P)$ as reachable in (N, M_{in}), if there exists $\sigma \in T^*$, s.t. $(N, M_{in}) \xrightarrow{\sigma} (N, M')$. The set of all reachable markings starting in (N, M_{in}) is denoted as $(N, M_{in}\rangle$.*

In the remainder of our work, we assume the existence of alignments between a Petri net and a log [6]. For this work, we assume that our log fits perfectly. For our approach, we need to remove silent transitions from alignments.

Definition 5 (Alignment). *Let $N = (P, T, F, A, l)$ be a Petri net and $L \subseteq \mathcal{U}_{ev}$ be a translucent event log. $T_\tau = \{t \in T \mid l(t) = \tau\}$ denotes the set of silent transitions. For a trace $\sigma \in \Sigma^L$, its perfectly fitting alignment on a Petri net N is denoted as $path(\sigma, N) \in T^*$. We link non-silent transitions to the corresponding events, i.e., $\pi^N_{trans}(\sigma_i) = (path(\sigma, N)\upharpoonright_{T \setminus T_\tau})_i$, for all $i \in \{1, \cdots, |\sigma|\}$.*

To access the behavior of a Petri net, we use reachability graphs. Each node in such a graph is a marking of a Petri net. Nodes are connected if a transition exists, s.t. firing the transition leads from one marking to the other.

Definition 6 (Reachability Graph). *Let $N = (P, T', F, A, l)$ be a labeled Petri net, with the initial marking $M_{in} \in \mathcal{B}(P)$. The reachability graph (RG) of the Petri net N is defined as $RG^{N, M_{in}} = (S, E)$ with $S = (N, M_{in}\rangle$ and $E = \{(M, t, M') \in S \times T \times S \mid \exists_{t \in T}\ (N, M) \xrightarrow{t} (N, M')\}$.*

The RG of the Petri net shown in Fig. 2a is displayed in Fig. 2b.

4 Translucent Precision

4.1 Log Behavior

We define events' prefixes using transitions based on alignments to capture the executed and enabled behavior.

Definition 7 (Executed and Enabled Behavior). *Let $L \subseteq \mathcal{U}_{ev}$ be a translucent event log, $N = (P, T, F, A, l)$ be a Petri net, and $T_\tau = \{t \in T \mid l(t) = \tau\}$ be the set of silent transitions. For a trace $\sigma \in \Sigma^L$, we define $\pi^N_{pref}(\sigma_i) = \pi^N_{trans}(pref_{i-1}(\sigma))$. The executed behavior for $e \in L$ at some point is defined as: $\pi^N_{prefact}(e) = \{\pi_{act}(e') \mid e' \in L\ \wedge \pi^N_{pref}(e) = \pi^N_{pref}(e')\}$ Similarly, we define the enabled behavior as: $\pi^N_{prefen}(e) = \bigcup_{\substack{e' \in L \\ \pi^N_{pref}(e) = \pi^N_{pref}(e')}} \pi_{en}(e')$.*

Given our example log shown in Table 1 and the Petri net depicted in Fig. 2a, $\pi^N_{prefact}(e_1) = \{a\}$, $\pi^N_{prefact}(e_2) = \{b, c\}$, $\pi^N_{prefact}(e_4) = \{e\}$, $\pi^N_{prefen}(e_4) = \{d, e\}$.

4.2 Model Behavior

When capturing model behavior, silent transitions provide a special challenge since their execution is not captured in the log. Moreover, executing them at a different point in time is often possible. As a result, we want to check if their execution enables other transitions, respectively, activities. Such activities could be captured as translucent activities. To do so, we first introduce τ-sequences. A τ-sequences is a sequence of transitions s.t. all, but the last transition is a silent transition. Given the Petri net depicted in Fig. 2a, possible τ-sequences are $\langle t_5, t_2\rangle$, $\langle t_5, t_3\rangle$, and $\langle t_5, t_4\rangle$. By using τ-sequences, we can transform an RG into a *direct RG*. In this process, we remove τ-transitions from the RG and establish connections between the start and end of these sequences.

Definition 8 (Direct RG). *Let $N = (P, T, F, A, l)$ be a Petri net, with its initial marking $M_{in} \in \mathcal{B}(P)$, and let Σ^N_τ be its set of τ-sequences. The Direct RG (DRG) of N is $DRG^{N,M_{in}} = (S, E)$ with $S = \langle N, M_{in}\rangle$ being the set of reachable markings and $E = E' \cup E_\tau$ s.t. $E' = \{(M, t, M') \in S \times T \times S \mid \exists_{t \in T \setminus T_\tau}\ (N, M) \xrightarrow{t} (N, M')\}$ and $E_\tau = \{(M, \sigma_{|\sigma|}, M')S \times T \times S \mid \exists_{\sigma \in \Sigma^N_\tau}\ (N, M) \xrightarrow{\sigma} (N, M')\}$.*

Figure 2c shows the DRG based on the previously shown RG (see Fig. 2b). Following the transition sequence $\langle t_1, t_2, t_3\rangle$ results in two markings: $[p_2, p_5]$ and $[p_4, p_5]$. Hence, following a transition sequence in the graph is not deterministic. To solve this problem, we simplify the graph using automata theory [17].

Definition 9 (Deterministic DRG). *Let N be a Petri net, with its initial marking $M_{in} \in \mathcal{B}(P)$, and $DRG^{N,M_{in}} = (S, E)$ be a DRG. The Deterministic DRG (DDRG) is a DRG, $DDRG^{N,M_{in}} = (S', E')$ s.t. $S' = \mathcal{P}(S)$ and $E' = \{(S_1, t, S_2) \in S' \times T \times S' \mid \exists_{t \in T}\ S_2 = \bigcup_{s_1 \in S_1} \{s_2 \mid (s_1, t, s_2) \in E\}\}$.*

The DDRG of the DRG depicted in Fig. 2c is shown in Fig. 2d. After making the replay deterministic, we want to access the enabled activities in a DDRG and, therefore, in the model. We use the states of the DDRG to do so.

Definition 10 (Enabled Activities in Model). *Let $N = (P, T, F, A, l)$ be a Petri net, with its initial marking $M_{in} \in \mathcal{B}(P)$, and $DDRG^{N,M_{in}} = (S, E)$ be the corresponding DDRG. For $s, s' \in S, t \in T$, if there exists an edge $(s, t, s') \in E$, we denote this with $s \xrightarrow{t} s'$. For $\sigma \in T^*$ and states $s_1, \ldots, s_{|\sigma|+1} \in S$, we denote $s_1 \xrightarrow{\sigma} (s_{|\sigma|+1})$, if $\forall_{1 \le i \le |\sigma|}\ s_i \xrightarrow{\sigma_i} s_{i+1}$. Let $L \subseteq \mathcal{U}_{ev}$ be a translucent event log. For $e \in L$ and $s \in S$ we define $\pi^{DDRG^{N,M_{in}}}_{state}(e) = s$ s.t. $\{M_{in}\} \xrightarrow{\pi^N_{pref}(e)} s$. Thus, $\pi^{DDRG^{N,M_{in}}}_{modelen}(e) = \{l(t) \mid \exists_{t \in T, s \in S}\ \pi^{DDRG^{N,M_{in}}}_{state}(e) \xrightarrow{t} s\}$.*

Given the example DDRG provided in Fig. 2d, $\pi^{DDRG^{N,M_{in}}}_{modelen}(e_1) = \{a\}$, $\pi^{DDRG^{N,M_{in}}}_{modelen}(e_2) = \{b, c\}$, $\pi^{DDRG^{N,M_{in}}}_{modelen}(e_3) = \{b, c\}$.

4.3 Computing Precision Scores

We first define a precision score, similar to escaping arcs, which does not consider enabled activities in the provided event log. Then, a score considering them.

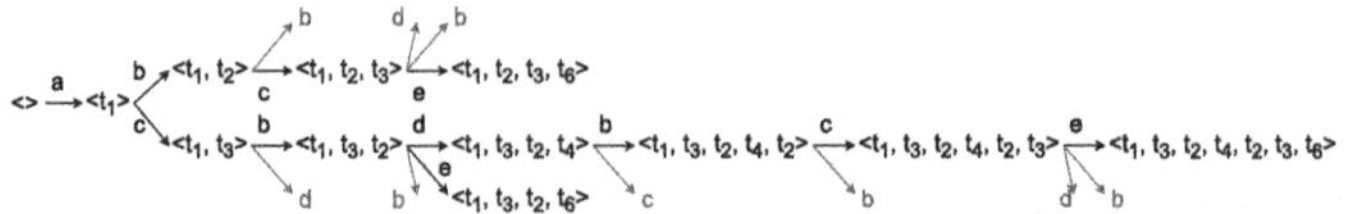

(a) Visualization of the computation of the precision score.

(b) Visualization of computing the translucent precision score.

Fig. 3. Visualization of the computation of the precision scores. Black arcs represent executed activities in the log, blue arcs represent enabled activities in the log, and red arcs represent activities enabled in the model but not in the log. (Color figure online)

Definition 11 (Precision Score). *Given a translucent event log $L \subseteq \mathcal{U}_{ev}$, a Petri net N with initial marking M_{in} and $DDRG^{N,M_{in}}$, we define precision as:*

$$prec(L, N) = \frac{1}{|L|} \cdot \sum_{e \in L} \frac{|\pi^{N}_{prefact}(e)|}{|\pi^{DDRG^{N,M_{in}}}_{modelen}(e)|}$$

Definition 12 (Translucent Precision Score). *Given a translucent event log $L \subseteq \mathcal{U}_{ev}$, a Petri net N with its initial marking M_{in} and its $DDRG^{N,M_{in}}$, we define the translucent precision score as follows:*

$$prec_t(L, N) = \frac{1}{|L|} \cdot \sum_{e \in L} \frac{|\pi^{N}_{prefen}(e) \cap \pi^{DDRG^{N,M_{in}}}_{modelen}(e)|}{|\pi^{DDRG^{N,M_{in}}}_{modelen}(e)|}$$

Note that we have to limit the numerator because activities that are not allowed in the model could be enabled. Illustrations for the methods are shown in Fig. 3. For our running example, we denote a precision score of 0.7 and a translucent precision score of roughly 0.78.

5 Evaluation

To evaluate our approach, we use the translucent event log shown in Table 1. Furthermore, we use the running example Petri net (Fig. 2a) and three additional Petri nets (Fig. 4) to evaluate whether the computed scores fit the meaning of

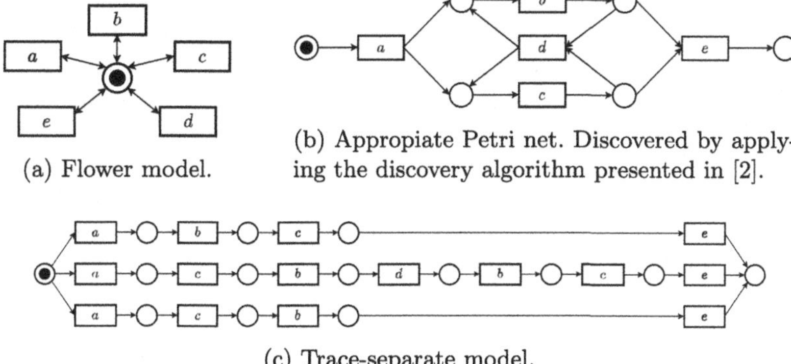

(a) Flower model.

(b) Appropiate Petri net. Discovered by applying the discovery algorithm presented in [2].

(c) Trace-separate model.

Fig. 4. Additional Petri nets.

Table 2. Precision and translucent precision scores for different models.

Model	Precision Score	Translucent Precision Score
Flower Model (Fig. 4a)	0.22	0.26
Example Model (Fig. 2a)	0.70	0.78
Appropriate Model (Fig. 4b)	0.90	1.00
Trace-separate Model (Fig. 4c)	1.00	1.00

precision. The results of our precision scores are displayed in Table 2. Both scores are low for the flower model. This shows that an imprecise model stays imprecise even when enabled activities are considered. The precision score from the example model suffers from b always being enabled after executing a and before executing e, thus allowing more behavior than caught in the log. Concerning the appropriate model, the traditional precision score suffers from the parallelism between b and c, and the choice between d and e. The translucent precision score considers enabled activities in the log, thus penalizing the afore-described behavior less. For the trace-separate model, we observe that both measurements yield a score of 1.0. In summary, the translucent precision method has the intuitive meaning of the traditional method, and considering enabled activities boosts the score for models that consider this information.

6 Conclusion

This paper presents the first notation and computational method of precision using translucent event logs. We showed that a well-established method can be extended to consider information on enabled activities. Also, we showed that our method still follows the natural understanding of precision by penalizing imprecise models. Furthermore, we showed that considering information on enabled activities is a valuable addition since process models that consider this knowledge get less penalized. Also, our method can handle duplicated transitions.

The approach we presented focuses on a fitting translucent event log. Hence, extending the approach to consider unfitting traces is valuable. Multiple methods exist for determining precision. Extending these techniques to consider information on enabled activities seems convenient. Moreover, methods for the other quality dimensions that consider translucent information should be introduced. When considering the different areas of process mining, techniques that consider the valuable information on enabled activities are needed.

References

1. van der Aalst, W.M.P.: Process Mining - Data Science in Action, 2nd edn. Springer, Heidelberg (2016). https://doi.org/10.1007/978-3-662-49851-4. ISBN 978-3-662-49850-7
2. van der Aalst, W.M.P.: Lucent process models and translucent event logs. Fundam. Informaticae **169**(1–2), 151–177 (2019). https://doi.org/10.3233/FI-2019-1842
3. van der Aalst, W.M.P., Adriansyah, A., van Dongen, B.F.: Replaying history on process models for conformance checking and performance analysis. WIREs Data Mining Knowl. Discov. **2**(2), 182–192 (2012). https://doi.org/10.1002/WIDM.1045
4. van der Aalst, W.M.P., Stahl, C.: Modeling Business Processes - A Petri Net-Oriented Approach. Cooperative Information Systems series. MIT Press, Cambridge (2011). ISBN 978-0-262-01538-7
5. Adams, J.N., van der Aalst, W.M.P.: Precision and fitness in object-centric process mining. In: ICPM. IEEE (2021). https://doi.org/10.1109/ICPM53251.2021.9576886
6. Adriansyah, A., van Dongen, B.F., van der Aalst, W.M.P.: Conformance checking using cost-based fitness analysis. In: EDOC, pp. 55–64. IEEE (2011). https://doi.org/10.1109/EDOC.2011.12
7. Adriansyah, A., Munoz-Gama, J., Carmona, J., van Dongen, B.F., van der Aalst, W.M.P.: Alignment based precision checking. In: La Rosa, M., Soffer, P. (eds.) BPM Workshops. LNBIP, vol. 132. Springer, Heidelberg (2012). https://doi.org/10.1007/978-3-642-36285-9_15
8. Adriansyah, A., Munoz-Gama, J., Carmona, J., van Dongen, B.F., van der Aalst, W.M.P.: Measuring precision of modeled behavior. Inf. Syst. E Bus. Manag. **13**(1), 37–67 (2015). https://doi.org/10.1007/S10257-014-0234-7
9. Beyel, H.H., van der Aalst, W.M.P.: Creating translucent event logs to improve process discovery. In: Montali, M., Senderovich, A., Weidlich, M. (eds.) ICPM Workshops. LNBIP, vol. 468. Springer, Cham (2022). https://doi.org/10.1007/978-3-031-27815-0_32
10. vanden Broucke, S.K.L.M., De Weerdt, J., Baesens, B., Vanthienen, J.: Improved artificial negative event generation to enhance process event logs. In: Ralyté, J., Franch, X., Brinkkemper, S., Wrycza, S. (eds.) CAiSE 2012. LNCS, vol. 7328, pp. 254–269. Springer, Heidelberg (2012). https://doi.org/10.1007/978-3-642-31095-9_17
11. vanden Broucke, S.K.L.M., De Weerdt, J., Vanthienen, J., Baesens, B.: Determining process model precision and generalization with weighted artificial negative events. IEEE **26**(8), 1877–1889 (2014). https://doi.org/10.1109/TKDE.2013.130

12. Carmona, J., van Dongen, B.F., Solti, A., Weidlich, M.: Conformance Checking - Relating Processes and Models. Springer, Cham (2018). https://doi.org/10.1007/978-3-319-99414-7. ISBN 978-3-319-99413-0
13. Chatain, T., Boltenhagen, M., Carmona, J.: Anti-alignments - measuring the precision of process models and event logs. Inf. Syst. **98**, 101708 (2021). https://doi.org/10.1016/J.IS.2020.101708
14. Chatain, T., Carmona, J.: Anti-alignments in conformance checking - the dark side of process models. In: Kordon, F., Moldt, D. (eds.) PETRI NETS. LNCS, vol. 9698. Springer, Cham (2016). https://doi.org/10.1007/978-3-319-39086-4_15
15. van Dongen, B.F., Carmona, J., Chatain, T.: A unified approach for measuring precision and generalization based on anti-alignments. In: La Rosa, M., Loos, P., Pastor, O. (eds.) BPM. LNCS, vol. 9850. Springer, Cham (2016). https://doi.org/10.1007/978-3-319-45348-4_3
16. Goedertier, S., Martens, D., Vanthienen, J., Baesens, B.: Robust process discovery with artificial negative events. J. Mach. Learn. Res. **10**, 1305–1340 (2009). https://doi.org/10.5555/1577069.1577113
17. Hopcroft, J.E., Motwani, R., Ullman, J.D.: Introduction to Automata Theory, Languages, and Computation, 2nd edn. Addison-Wesley series in computer science. Addison-Wesley-Longman (2001). ISBN 978-0-201-44124-6
18. Leemans, S.J.J., Polyvyanyy, A.: Stochastic-aware precision and recall measures for conformance checking in process mining. Inf. Syst. **115**, 102197 (2023). https://doi.org/10.1016/J.IS.2023.102197
19. Munoz-Gama, J., Carmona, J.: A fresh look at precision in process conformance. In: Hull, R., Mendling, J., Tai, S. (eds.) BPM. LNCS, vol. 6336. Springer, Heidelberg (2010). https://doi.org/10.1007/978-3-642-15618-2_16
20. De Weerdt, J., De Backer, M., Vanthienen, J., Baesens, B.: A robust F-measure for evaluating discovered process models. In: CIDM. IEEE (2011). https://doi.org/10.1109/CIDM.2011.5949428

IPMD: Intentional Process Model Discovery from Event Logs

Ramona Elali[1(✉)], Elena Kornyshova[2], Rébecca Deneckère[1], and Camille Salinesi[1]

[1] Paris 1 Panthéon Sorbonne, Paris, France
{ramona.elali,rebecca.deneckere,camille.salinesi}@univ-paris1.fr
[2] Conservatoire National des Arts et Métiers, Paris, France
elena.kornyshova@cnam.fr

Abstract. Intention Mining is a crucial aspect of understanding human behavior. It focuses on uncovering the underlying hidden intentions and goals that guide individuals in their activities. We propose the approach IPMD (Intentional Process Model Discovery) that combines Frequent Pattern Mining, Large Language Model, and Process Mining to construct intentional process models that capture the human strategies inherited from his decision-making and activity execution. This combination aims to identify recurrent sequences of actions revealing the strategies (recurring patterns of activities), that users commonly apply to fulfill their intentions. These patterns are used to construct an intentional process model that follows the MAP formalism based on strategy discovery.

Keywords: Intention Mining · Intentional Process Model · Frequent Pattern Mining · Process Mining · Large Language Model

1 Introduction

Each human being acts to complete an initial target. This target can be reached through a different set of activities. Hence, activity logs are the main source of information to construct a process model [1]. A standard process model is activity-oriented and lacks an abstract overview that describes the main purpose of an executing activity. Intention Mining (IM) has become an essential research field where it is widely now applicable to different domains of application using different techniques [2]. IM came across to address the intentional process models. Thus, we believe that IM can provide us with a higher quality of recommendations [5] since it focuses on the intentional part of processes that is closer to the user's thinking and goals. The main objective of IM is to know the users' goals and purpose while using the system by discovering the intentional process models that reflect the user's behavior and strategy. In addition, the actual processes enacted by the users differ from the established business processes [8]. The main reason for this difference is that the user doesn't rely only on the prescribed processes to complete his daily activities, but he also uses his intentions and strategies to complete a certain task.

[*] https://github.com/ramonaelally/IntentionalProcessModelDiscovery

J. Araújo et al. (Eds.): RCIS 2024, LNBIP 514, pp. 38–46, 2024.
https://doi.org/10.1007/978-3-031-59468-7_5

IM focuses on the why rather than what or when [13]. Hence, this work focuses on the Intentional Process Model Discovery (IPMD) using MAP formalism [13]. Event logs lack the necessary information to deduce the user intention behind each activity. It is not straightforward to use an event log to construct an intentional process model. The existing work [9] uses a top-down approach. On the contrary, our approach is bottom-up as we use the event logs to identify the distinct elements of the MAP formalism to construct the intentional process model in a semi-automatic way. Our approach IPMD combines different techniques coming from different fields, such as Frequent Pattern Mining (FPM) [13], Process Mining (PM) [1], and Large Language Models (LLM) [7] to achieve this goal. In the following section, we describe the IPMD approach with its application and we conclude in Sect. 3.

2 Intentional Process Model Discovery (IPMD)

Human beings behave according to their intentions and tend to not stick to a fixed schedule or a routine to perform their activities. Their behavior is variable and depends on multiple constraints. Normally, each intention is achieved by a set of activities representing a specific strategy. Hence, we consider these concepts on three different levels: the intentional level (intention), the operational level (activity), and the strategic level (strategy). For instance, a person can fulfill a specific intention through different strategies, and a strategy can be achieved by different sets of activities. Understanding and mining these intentions are important for constructing precise and significant process models. For that purpose, our main research question for this work is *How to define an IM algorithm that uses logs as sources to build an intentional process model?* A solid theoretical foundation was built for this topic by conducting a systematic literature review and a research agenda on IM [6] using the comparative framework presented in [2], in addition to a literature review and research agenda on Intentional Process Engineering [3].

In this current work, we are interested in discovering intentional process models based on MAP formalism [13]. By relying on the results of the systematic literature review, we can notice that only a few works target intentional process modeling [8, 9, 13] and specifically rely on MAP formalism to present the intentional process model using a top-down approach. Our purpose is the discovery of intentional process models. Previous work used a top-down approach [9] that depends on interviews, observations, meta-models, and existing models to construct intentional process models, which need a huge effort between different parties. In our approach, we follow a bottom-up approach that relies only on the activity logs and the meta-model as initial input.

The MAP model is a labeled directed graph with intentions as nodes and strategies as edges that connect the intentions [13]. The MAP model comprises sections as shown in Fig. 1 and Fig. 2. The sections are composed of triplets: the source intention, the target intention, and the strategy $< I_{source}, I_{target}, S_{source\text{-}target} >$ [13]. In MAP, a strategy is defined as an approach or a manner to achieve an intention. Each section can be related to a group of activities (in sequence, in parallel, represented as a complex activity-oriented process model, and so on). As shown in Fig. 2, we consider three levels: intentional, strategic, and operational. MAP is presented in the intentional aspect of a process. The sections (a strategy linking two intentions) are on the strategic level, whereas the activities are on the operational level. The hierarchical structure depicted in Fig. 2 (intentional to strategic to operational) not only improves the granularity of understanding but also gives a comprehensive and full view of the interdependencies among the various parts. Generally, the set of activities that are done by a human completes a section that achieves a strategy that tends to fulfill a specific intention, as described in Fig. 1 and Fig. 2. In a top-down approach, using only the intentional and the operational layer is enough, but in a bottom-up approach, it is required to identify the intermediate level to be able to generate the intentional model. That's why we decomposed the intentional layer into two levels: intentional and strategic.

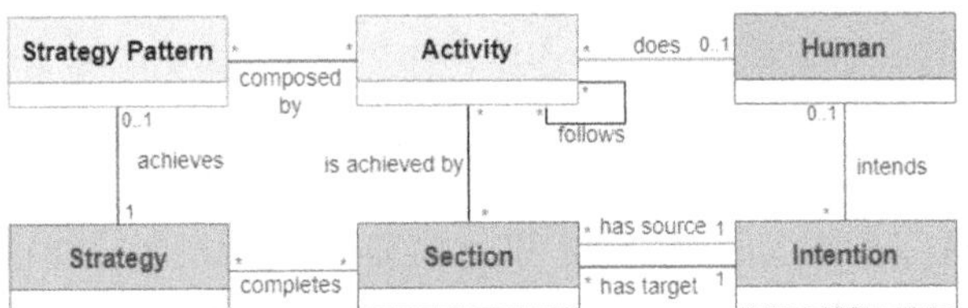

Fig. 1. IPMD Approach metamodel.

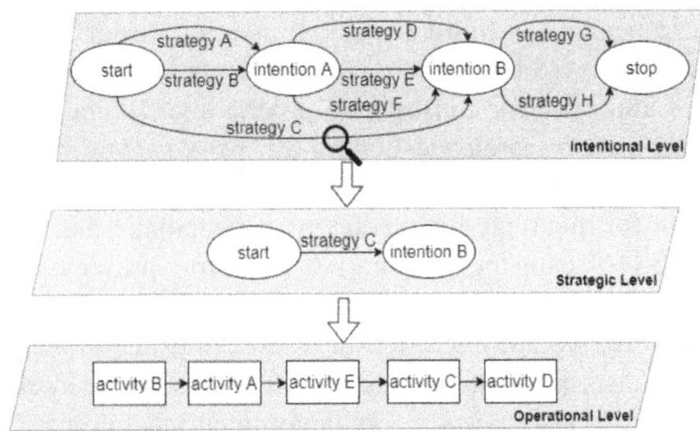

Fig. 2. MAP and the Hierarchical Structure Overview.

To build the intentional process model it would be crucial to identify the strategy patterns that interlink the activities with the intentions. Our proposal IPMD combines three techniques: FPM allows us to discover the strategy patterns from the activities (going from the operational level to the strategic level), PM is used to construct the process model based on the strategy patterns, and LLM is utilized to name the strategies and the intentions (to go from the strategic level to the intentional level). In the original meta-model of MAP [13], the strategic and the operational layers are not represented as specific concepts (activities are embedded inside the concept of the section and the strategy is only connected to the section). Therefore, we adjusted the meta-model to fit our needs. We added the activity concept to the diagram as indicated in Fig. 1. In addition, we added the strategy pattern, because, in our bottom-up approach, the activity log is the single input that we have (except from the meta-model definition), to go to the strategic level. We do not have a direct link between the strategy and the activity; hence we need to identify the strategy patterns that can be fetched from the initial activity log to reach the strategic level. That's why adding Activity and Strategy Pattern concepts to the MAP meta-model was crucial to elaborate our approach. The activity log is a starting point to construct the strategy pattern log which will be followed by the intentional process model construction. Our approach, which is described in the following with its application, is composed of two main phases. In this work, we have used the BP-Meets dataset [10] that describes the daily habits of an individual living in a home. In the following, we will present the application of the proposed approach.

2.1 Phase 1: Strategy Pattern Log Construction

Figure 3 represents Phase 1 of the approach proposed in this work. A section (a strategy for achieving an intention) is composed of a set of activities on the operational level. Hence, as a first step, since the activity is not directly linked to the strategy or the intention, we must identify the strategy patterns based on the activity log. Then, after identifying the different groups of activities, we should give each group a meaningful name that accurately represents the strategy pattern. The final step of this phase requires generating a strategy pattern log.

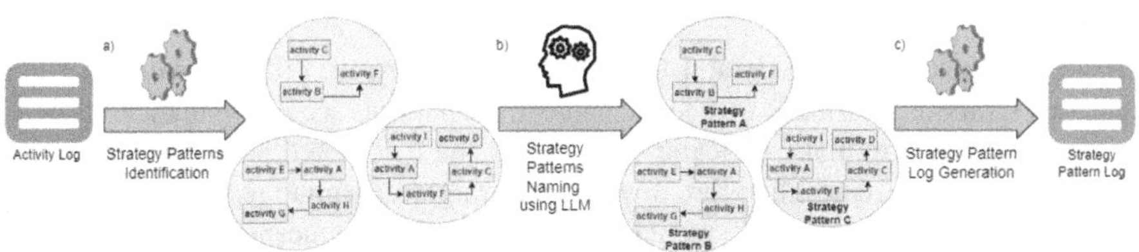

Fig. 3. Phase 1 of the IPMD Approach.

Strategy Patterns Identification. The identification of strategies in IM plays a key role as it offers a deeper knowledge of the regular tactics that individuals employ to fulfill their intentions. The set of activities that occurred together for a specific strategy to achieve an intention can be considered as a pattern that is being repeated while representing an intended behavior. Hence, discovering strategies is like uncovering recurrent patterns within the complicated structure of human behavior. Hence, taking advantage of FPM techniques helps find the strategy patterns (the set of activities that tend to occur together). Then, by analyzing the identified frequent patterns we can effectively extract and outline the strategies commonly used to achieve specific intentions. FPM techniques can unveil the strategic plans that guide human actions. Different FPM techniques exist [14], that allow us to discover interesting and meaningful patterns and associations within an existing dataset. However, each algorithm has its downside effects and advantages. PrefixSpan addresses the challenges of mining frequent patterns in large sequences. It is a sequential pattern mining algorithm that performs pattern mining incrementally without generating candidate sequences explicitly. It is effective for handling large-scale sequential data. In this work, as we are interested in the identification of sequences of activities from an activity log where the sequence tends to be large, PrefixSpan fits our needs because of its memory efficiency. In our case study, at first, we did a manual grouping of the activities according to their type after analyzing the activity log to obtain a reference set (to check later if our natural way of thinking was aligned with the automatically discovered patterns). Then, we tried to identify those groups automatically using PrefixSpan. The objective was to find the patterns (set of activities that tend to happen together). Each pattern represents a strategy that is employed by the user to achieve his intention. To apply the PrefixSpan algorithm we need to make a pre-treatment of the data log. After applying the PrefixSpan algorithm, we obtained the frequent sequential patterns that represent activities that frequently occur together in the same order. We obtained 15 patterns. We illustrate a selection of these patterns in Table 1.

Table 1. PrefixSpan Strategy Patterns Sample

Patterns	
Pattern 1	go_wardrobe, get_clothes, change_clothes, go_bathtub, have_bath, go_bathroom_sink, brush_teeth, go_bed, sleep_in_bed
Pattern 2	go_wardrobe, change_clothes, go_shoe_shelf, dress_up_outdoor

Strategy Patterns Naming Using LLM. After identifying the different patterns using the pattern-mining algorithm, we need to name those patterns according to their purpose to ease the use and to make it clearer to the user. The discovered patterns are initially groups of named activities therefore it would be straightforward to use LLM from Natural Language Processing (NLP) techniques to extract the real meaning that describes a set of activities (since the activities are initially named according to their actual function). Pointing out that a set of activities aims to achieve a specific intention. Different language models and techniques already exist that can facilitate this task and GPT-based model [12] is one of them. GPT models can capture, understand, and maintain the contextual information in a given request or query. They can generate coherent and contextually relevant responses. In this work, we integrate our work with GPT models using OpenAI API [11] to accomplish this task which consists of giving a meaningful name to each pattern separately. A list of meaningful names coming from the GPT model is constructed, and then the best name is chosen, either automatically (by choosing the top name in the list), or manually (directly by the process engineer). We developed a Python script* that connects and uses OpenAi API to give a meaningful name to each pattern dynamically. We are using "gpt-3.5-turbo" [11] as LLM because it has a lower cost and improved performance. We executed the API iteratively on each pattern 10 times to get a different list of names. We saved these names inside a file to allow the user to choose the suitable name for each pattern as in Table 2 (The Count column represents the number of times a name was suggested through the iterations).

Strategy Pattern Log Generation. After naming the different strategy patterns, we need to construct the log to move a layer up from the operational level to the strategic level. The initial activity log will be used as a base to construct this strategy pattern log.

Table 2. Strategy Patterns Naming Suggestions Example

Patterns	Suggestions	Count
Pattern 1	Bedtime Routine	4
	Daily Bedtime Routine	6
Pattern 2	Getting Ready	2
	Clothing Routine	4
	Getting Dressed	4

We need to replace each group of activities with a suitable strategy pattern relying on the previously discovered named patterns. We keep some of the initial log attributes, such as the resource id. We also add new attributes, such as two timestamps (Start Timestamp and End Timestamp, which represent the start and end of the set of activities of a specific pattern) instead of a timestamp for each record. Hence, a strategy pattern log will be constructed as a result. We developed a C# algorithm* that achieves this requested task to obtain a strategy pattern log (Table 3 represents an example of the generated log).

Table 3. Strategy Pattern Log Example

Id	Name	Resource Id	Start Timestamp	End Timestamp
531	Bedtime Routine	1	3/25/2020 00:10:00	3/25/2020 6:54:00
574	Clothing Routine	1	3/26/2020 9:15:00	3/26/2020 9:25:00

2.2 Phase 2: Intentional Process Model Construction

Figure 4 represents Phase 2 of the IPMD approach. It consists of several steps. The main goal of this phase is to identify the intentions achieved by each of the discovered strategy patterns, which is a crucial step in building the intentional process model. To discover the intentions, we need to use PM and LLM. In the following, we describe each step.

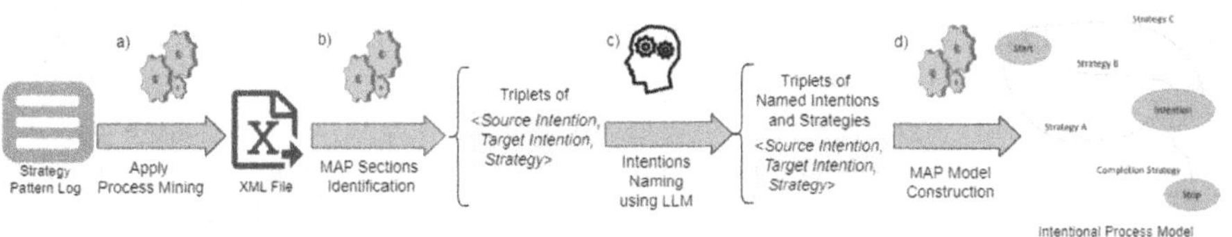

Fig. 4. Phase 2 of the IPMD Approach.

Applying Process Mining and Exporting XML File. We use the strategy pattern log to build a process model that will represent the strategy patterns instead of the activities.

We use PM techniques to build the strategy pattern process model as an intermediate step before constructing the intentional process model. We need to export the strategy patterns process model into XML format to extract the essential information that will be used as input in the next step to identify the sections' elements. PM is used to discover, check conformance, and enhance business processes, and recommendations by analyzing event logs recorded during the real-time execution of these processes [1]. Different algorithms can be used in each technique and through this step, we are interested in the discovery techniques. In discovery techniques, we have Alpha Algorithm, Heuristic Mining, Fuzzy Mining, etc. We will not go inside an explanation of all the different algorithms of PM that exist in [1]. There are a lot of them that allow us to obtain a process model that can be expressed as an XML file to be able to go further. We used Disco [4] as a tool for PM because it is free, simple, and easy to use. We exported the strategy patterns process model into an XML format which will be used as input in the next step.

Sections Identification. Sections are an important element of MAP formalism. Each section comprises a source intention, a target intention, and a strategy. To construct the requested intentional process model identifying the sections is a necessary step. The extracted information from the XML file from the previous step will be used as input for a developed algorithm* in C# that will transform it into a set of sections (Source Intention, Target Intention, and Strategy) as shown in Table 4.

Table 4. Sections Examples

Sections
< Intention8, Intention0, Bedtime Routine >
< Intention0, Intention3, Clothing Routine >

Intentions Naming Using LLM. After identifying the sections, each section is presented with an anonymized name of the intention source and target. Hence, it is required to name the intentions following the same method applied to the strategy patterns. In this step, we already have the strategy pattern names, so we are relying on them to define the intention names of each section while starting from a Start Intention and then using each section's details to describe the next linked section. In the same way, as in the previous phase, a set of names will be obtained through the GPT model, and the selection of the exact name can be done automatically or manually.

We developed a Python script* that uses OpenAI API. It names dynamically the intentions while using each section's information to define the intention name in the next one. As an illustration, the section < Intention8, Intention0, Bedtime Routine >. We will use this information to extract the target intention name "Intention0" from GPT model. The intention name would be "Sleep". Then, we will use this discovered information to identify the intention in the next linked sections < Intention0, Intention3, Clothing Routine >. Then we obtain for Intention3 to be "Dress" (<Sleep, Dress, Clothing Routine >).

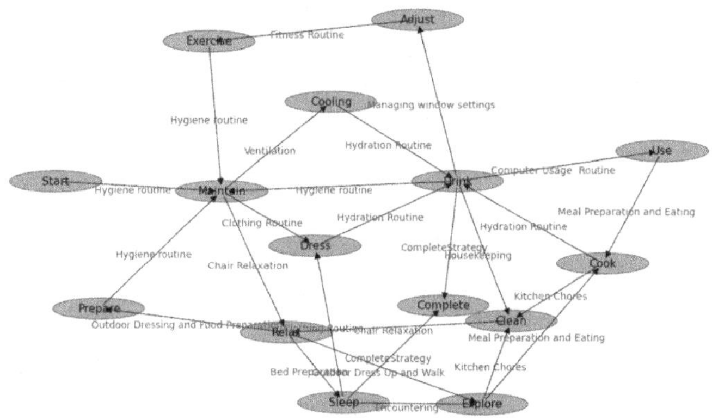

Fig. 5. The Intentional Process Model (MAP).

Intentional Model Construction. Now that the sections have been correctly designed, the MAP model can be derived automatically. We developed a Python script* that takes the set of sections as input and constructs the intentional process model graphically (Fig. 5).

3 Conclusion

To conclude, this work has presented the IPMD approach. It showcased the success of constructing intentional process models based on user activity logs and MAP formalism. By exploiting techniques such as Frequent Pattern Mining, Process Mining, and Large Language Modeling, we've presented the extraction of meaningful strategy patterns and intentions from activity logs. Acknowledging the dynamic nature of human behavior is influenced by various external and environmental factors such as time of day, location, user preferences, etc., our future works will focus on integrating contextual information with IPMD.

References

1. Aalst, W.M.P.: Process Mining: Data Science in Action. Springer, Heidelberg (2016). ISBN: 978–3–662–49850–7. https://doi.org/10.1007/978-3-662-49851-4
2. Déneckère, R., Kornyshova, E., Hug, C.: A framework for comparative analysis of intention mining approaches (2021).https://doi.org/10.1007/978-3-030-75018-3_2
3. Déneckère, R., Kornyshova, E., Elali, R.: Intentional Process Engineering: Literature Review and Research Agenda (2023)
4. Disco. https://fluxicon.com/disco/
5. Elali, R.: An intention mining approach using ontology for contextual recommendations. Proceedings of the Doctoral Consortium Papers Presented at the 33rd International Conference on Advanced Information Systems Engineering (CAiSE 2021), Melbourne, Australia, June 28 - July 2, 2021. CEUR Workshop Proceedings, vol. 2906, pp. 69–78. CEUR-WS.org (2021)

6. Elali, R., Déneckère, R., Kornyshova, E.: Intention Mining: a systematic literature review and research agenda (2024)
7. Kalyan, K.S.: A survey of GPT-3 family large language models including ChatGPT and GPT-4. Nat. Lang. Process. J. **6**, 100048 (2024). ISSN 2949–7191
8. Khodabandelou, G.: Contextual recommendations using intention mining on process traces, Doctoral consortium paper. In: International Conference on Research Challenges in Information Science, RCIS (2013)
9. Khodabandelou, G., Hug, C., Deneckère, R., Salinesi, C.: Supervised intentional process models discovery using hidden markov models. In: International Conference on Research Challenges in Information Science, RCIS (2013)
10. Koschmider, A., Leotta, F., Serral, E., Torres, V.: BP-Meets-IoT 2021 Challenge Dataset (2021)
11. OpenAI API. https://platform.openai.com/overview
12. Radford, A., Narasimhan, K., Salimans, T., Sutskever, I.: Improving language understanding by generative pre-training (2018)
13. Rolland, C., Prakash, N., Benjamen, A.: A multi-model view of process modelling. Requirements Eng. **4**(4), 169–187 (1999). https://doi.org/10.1007/s007660050018
14. Zaki, M., Meira, W., Jr.: Data Mining and Analysis: Fundamental Concepts and Algorithms. Cambridge University Press, Cambridge (2014)

Forensic-Ready Analysis Suite: A Tool Support for Forensic-Ready Software Systems Design

Lukas Daubner[1(✉)], Sofija Maksović[1], Raimundas Matulevičius[2], Barbora Buhnova[1], and Tomáš Sedláček[1]

[1] Masaryk University, Brno, Czechia
{daubner,maksovic,buhnova,tomas.sedlacek}@mail.muni.cz
[2] University of Tartu, Tartu, Estonia
raimundas.matulevicius@ut.ee

Abstract. Forensic-ready software systems integrate preparedness for digital forensic investigation into their design. It includes ensuring the production of potential evidence with sufficient coverage and quality to improve the odds of successful investigation or admissibility. However, the design of such software systems is challenging without in-depth forensic readiness expertise. Thus, this paper presents a tool suite to help the designer. It includes a graphical editor for creating system models in BPMN4FRSS notation, an extended BPMN with forensic readiness constructs, and an analyser utilising Z3 solver for satisfiability checking of formulas derived from the models. It verifies the models' validity, provides targeted hints to enhance forensic readiness capabilities, and allows for what-if analysis of potential evidence quality.

Keywords: Forensic Readiness · Forensic-by-Design · Forensic-Ready Software Systems · Z3 Solver · BPMN · Modelling

1 Introduction

Forensic readiness is a set of precautions and activities aiming to facilitate prompt and successful investigation [20,23] of an incident. In some sense, forensic readiness prepares for the situation where preventive measures fail. The investigations help to establish the root cause and culprits, discover the extent of damage and prevent future incidents [2,3,9]. Furthermore, forensic readiness supports dispute resolution and proving of due diligence [20].

The software systems can be designed to integrate forensic readiness [10], called forensic-ready or forensic-by-design. They proactively collect potential digital evidence[1] to be used in the investigation and soundly conduct forensic processes [16]. However, designing such systems to meet the needs is challenging.

[1] Note the difference: potential digital evidence – potentially useable for future investigation, and digital evidence – used to satisfy or refute the investigation hypothesis.

J. Araújo et al. (Eds.): RCIS 2024, LNBIP 514, pp. 47–55, 2024.
https://doi.org/10.1007/978-3-031-59468-7_6

One possible approach is to employ risk management to assess the system and formulate specific, verifiable requirements [7,13]. Furthermore, a model-based approach can complement it. For example, BPMN for Forensic-Ready Software Systems (BPMN4FRSS) [4,5] models capture the requirements and derive metrics for assessment. Still, it needs to be supported with tools and more in-depth analyses to facilitate adopting the model-based design approach.

This paper presents the Forensic-Ready Analysis Suite (FREAS), a tool addressing the research question: **How to support the modelling and analysis of forensic-ready software systems?** It integrates two results of theses: an editor for BPMN4FRSS models [21] and an analyser for the models [12]. Additionally, the analyser is extended with new checks. The tool allows for modelling the forensic-ready software system in BPMN4FRSS notation and analysis of the models based on formula satisfiability checking using the Z3 solver [1,15]. It checks the model's correctness, suggests hints to improve it, and explores what-if scenarios to determine the quality of potential evidence within the system.

2 Related Work

Multiple modelling approaches focus on designing forensic-ready systems, each focusing on different use cases. These are enhanced Business Process and Notation (BPMN) [6], UML Sequence [19], and UML Activity [22] diagrams. The model-based analysis techniques and tools are well-recognised. An example of a secure design is UMLsec [11], a UML extension supporting the analysis of models regarding security. For privacy design, there is PE-BPMN [18], a BPMN extension aimed at analysing privacy leakages using the PLEAK tool [24].

The utilisation of Z3 solver in software analysis is also well-known. It is one of the techniques employed in the aforementioned PLEAK tool [24]. For security context, it was used for security trade-off balancing [17] and to locate inconsistencies between firewall and security policies [25].

While tools are available for security analysis of BPMN models, the forensic readiness aspects are not addressed. Thus, a tool to fill this gap is needed. Furthermore, the Z3 solver is not common in state-of-the-art BPMN tools. In this paper, we incorporate it to enable the assessment of the designed model.

3 Forensic-Ready Analysis Suite

The Forensic-Ready Analysis Suite (FREAS) is an open-source[2] tool that aims to assist with forensic-ready software design in two areas. First, to bridge the gap of forensic readiness expertise, i.e., support the designer in introducing correct baseline forensic readiness into the system without needing a deep knowledge of the subject. Second, to support the forensic Forensic-Ready Information Systems Security Risk Management (FR-ISSRM) [4] in enabling a more detailed evaluation of the forensic readiness scenarios and proposed treatments.

[2] The documentation is available at: https://freas-tools.github.io/wiki/.

FREAS is composed of a Modeller and a Rule-Based Analyser. Prototype implementations of the components were a topic of theses [12,21]. The version presented in this paper integrates and extends them.

Modeller is a front-end web application for creating and editing BPMN4FRSS diagrams, which can be exported as models. It is based on the *bpmn-js* javascript library, a part of the bpmn.io project, for rendering and editing BPMN diagrams. The library is extended with BPMN4FRSS specification [6], which includes plugging the extended model syntax, editor behaviour and diagram renderer to support BPMN4FRSS modelling. The extension is encapsulated in a reusable *freas-bpmn4frss-library* library and integrated into a React-based application that wires the functionality and provides a GUI for the analysis.

Rule-Based Analyser encapsulates the analysis of BPMN4FRSS models, using Z3 solver. It defines three analysis types, corresponding to three sets of rules, defined as various first-order logic formulas in combination with background theories [15]. These are **Validity**, **Hint**, and **Evidence Quality**. The rules are constructed from the facts parsed from the model and formulas specifying the rule itself. For example, a validity rule: *All evidence sources have a 'Produces' association.* However, as Z3 checks for satisfiability rather than validity, the rule needs to be negated: *There is an evidence source not having a 'Produces' association.* That way, the resulting satisfiable solution contains the elements breaking the rule. Otherwise, the formulas are unsatisfiable – the rule is not broken.

4 Demonstration: Automated Valet Parking System

This section explains the capabilities of FREAS and demonstrates how it can be used to model and analyse an Automated Valet Parking (AVP) system[3].

4.1 System Description and Modelling

The AVP system offers a service enabling the user to leave an autonomous vehicle in a drop-off area to park it automatically. Originally represented as BPMN process model [7,8], it was enhanced with BPMN4FRSS constructs to add forensic readiness. The system involves three separate entities:

Autonomous Vehicle (AV): Representing the vehicle communicating with the Parking Service Provider to obtain a parking permit. Here, it is considered non-cooperative (i.e., potential evidence is not expected to be available or reliable).

Parking Service Provider (PSP): Representing a contact point of the cloud-based parking service, orchestrating search, reservation, and parking permit delivery. It is cooperative (i.e., potential evidence is expected to be available).

[3] Code, models, and a video demo are available at: https://doi.org/10.58126/bcxs-cr23.

Parking Lot Terminal (PLT): Representing an edge device physically located at the parking lot, thus with easy physical access. It controls access to the parking lot by issuing and checking the parking permits. The entity is cooperative.

This demonstration focuses on an AV getting a valid parking permit to park at the parking lot later. The process is as follows: The AV sends a Parking request to PSP, which checks for parking space availability at the PLT. If available, PSP generates a parking reservation and sends it to PLT, which generates a parking permit. The permit is then sent back through PSP to the requesting AV.

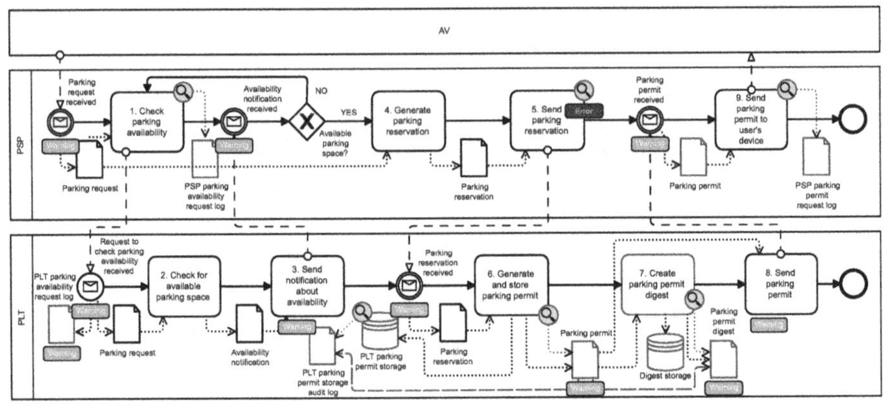

Fig. 1. AVP: Issuing a Permit – Validity and Hint Analysis

First, the designer uses the Modeller to draft an initial design of the AVP system as a BPMN diagram (see Fig. 1). It contains three *Pools* (AV, PSP, PLT) – the three entities and contexts communicating with *Message Flows*. The *Pools* contain a combination of *Tasks* (rectangles) representing atomic activities, *Events* (circles), and *Gateways* (diamonds) to describe a process from the start to the end. These are supplemented by *Data Objects* (documents) representing data and *Data Stores* (cylinders) representing persistent storage.

Then, the BPMN4FRSS extension is applied by marking *Data Objects* as *Potential Evidence* (green Data Object) with its *Evidence Source* (magnifier) as a point of their origin and *Evidence Stores* (green Data Stores) as their persistent storage. Further constructs are specialised forensic readiness-specific *Tasks* (in green), and *Evidence Associations* (green arrows) representing a relationship (a correlation) between *Potential Evidence.*

As the designer focuses on forensic readiness, the AV Pool is collapsed due to its non-cooperativity [6], as its data cannot be relied upon. They also added several Evidence Sources, producing Potential Evidence as logs. Furthermore, the Parking permit itself is marked as Potential Evidence (proving the execution of the issuing process) and stored on PLT. As an additional measure, a digest (i.e., hash) of the Parking permit is computed and stored as proof of its integrity [5].

4.2 Validity and Hint Analysis

To ensure the correctness of the model and receive feedback, the designer performs Validity and Hint Analysis on the created model by pressing a button in the GUI. The **Validity Analysis** stems from the BPMN4FRSS model restrictions [6], translated to first-order logic formulas with background theories. They point out syntactical or semantical errors. Thus, it reports on incorrectly constructed parts of the model or parts that are well constructed but do not make sense in what a BPMN4FRSS model represents [6]. On the other hand, **Hint Analysis** checks formulas that identify where the model can be improved (e.g., by introducing new evidence sources [7]) or that a forensic readiness control is not designed effectively. Thus, they give the designer feedback, pointing out parts that should be re-examined and what they should consider.

The initial AVP system design is subjected to Validity and Hint Analysis as highlighted in Fig. 1. FREAS displays a red label "Error" denoting an element with validity error, and an orange label "Warning" denoting a hint regarding the element. The labels show detailed messages on hover. The results are as follows:

- **Evidence Source must have Potential Evidence** (Validity Error) – Evidence Source not producing Potential Evidence is meaningless.
- **Message Flow source and destination should have Evidence Source** (Hint) – The analyser hints at having Potential Evidence on a communication interface, so utilisation of the particular channel can be confirmed.
- **Potential Evidence should have Evidence Source defined** (Hint) – Potential Evidence should have a point of origin defined. The lack of it points to a gap in the design.
- **Potential Evidence should be stored in a different Evidence Context than its Proof** (Hint) – Usage of hashing is a known practice to prove integrity [20]. However, keeping both in the same context might not bring the desired assurance, as it does not protect against whole context compromise.

Following the feedback, the designer resolved the validity errors and implemented the chosen hints, producing a second iteration, shown in Fig. 2. Specifically, the PSP parking reservation request log is added, and Evidence Source is defined for the PLT parking availability request log. Furthermore, Digest storage is moved to PSP; thus, the Parking permit digest is also sent to it. The remaining hints are acknowledged by the designer and left unimplemented.

4.3 Evidence Quality Analysis

With a valid BPMN4FRSS model with a baseline forensic readiness included, the designer moves to check whether the modelled system can post-mortem detect a security risk occurrence. In other words, whether there would be usable evidence within the system to prove the impact of the risk (a forensic readiness goal [4]).

As the risk impacts and compromises a part of a system, it also affects the related potential evidence. If unaccounted for, it might lead to incorrect conclusions and interfere with the forensic investigation. Thus, the analysis indirectly refers to the evidentiary value, supporting admissibility [14]. The higher value indicates higher confidence in the resulting evidence, making it more challenging to dispute. Thus, it reflects the corroborability of potential evidence [5].

The Evidence Quality Analysis defines formulas that are satisfied with Evidence Stores that hold data and can help detect inconsistencies in the Potential Evidence. If there are none, the model should be enhanced as needed.

First, the designer conducts the security risk management [13] to identify the risks. Then, it is further iterated in a forensic-ready risk management process [4]. Specifically, the risks considered for the AVP system are as follows:

1. A malicious insider injects a Parking permit into the PLT parking permit storage due to their permissions. Leading to loss of Parking permit integrity.
2. An attacker fabricates a Parking reservation and sends it to the PLT interface due to missing access control. Leading to loss of Parking permit integrity.

Using the analysis, the designer put what-if queries on the model. They chooses the affected elements to be considered compromised and starts the analysis individually. Those are PLT parking permit storage and Parking reservation received, respectively. The Evidence Quality Analysis checks the relations between the Potential Evidence and their place in the process to determine which potential evidence could be used to spot a discrepancy. It could be missing or extra data or a value difference. Notably, the analysis reports where the potential evidence can be found and where it can be retrieved (Evidence Store).

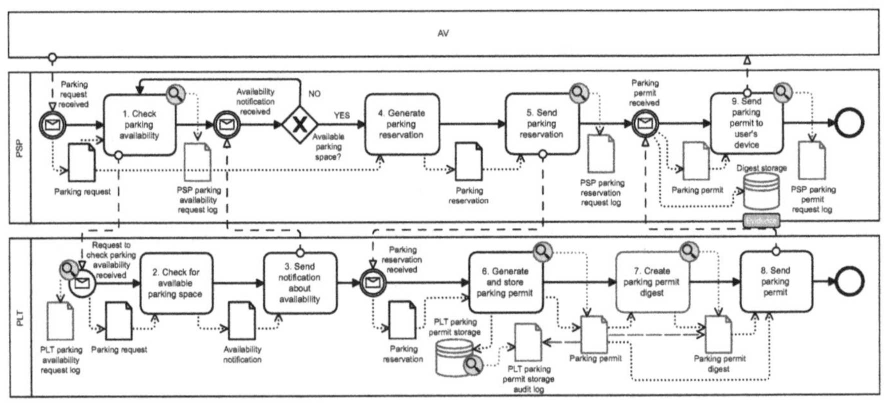

Fig. 2. AVP: Issuing a Permit – Evidence Quality Analysis

In the first case, the compromised PLT parking permit storage can be determined by Parking permit digest, stored in Digest storage. An out-of-process injection would not create the digest or store it in PSP. As a result, the discrepancy would be the non-existence of a Parking permit digest. Thus, the element is marked accordingly as illustrated in Fig. 2.

However, in the second case, the analyser does not find a suitable store, prompting the designer to adapt the design. The issue could be resolved by adding new a Evidence Store on the PSP side to store the corresponding Parking reservation or the logs (PSP parking reservation request log). They would also need to define the relationships between the potential evidence in the model.

5 Conclusion

This paper presented a modelling and analysis tool for BPMN4FRSS models, addressing the research question: **How to support the modelling and analysis of forensic-ready software systems?** First, the Modeller assists in capturing a forensic-ready software system as a BPMN4FRSS model, enforcing the basic syntactical rules of BPMN4FRSS. Then, based on formal satisfiability checking, the Rule-Based Analyser gives the designer feedback on the model, finds errors in the design and proposes enhancements. It also allows the designer to query the model based on the risks to evaluate available potential evidence in forensic readiness scenarios. Together, the tool facilitates the designing of forensic-ready software systems for a broader audience, helping to bridge the gap of needed forensic readiness expertise. We report the following learnt lessons:

- The Modeller helps the designer record the information regarding forensic readiness, including potential evidence, their relations, sources, and storage.
- The Rule-Based Analyser guides the designer in thinking about the relations of potential evidence, how it is stored, and its contribution concerning risks.
- The FREAS tool facilitates the addition of forensic readiness to software systems at the business logic level.

Future Work: The tool suite opens up new possibilities for forensic-ready software systems design to be addressed in future work. Additional rules and Z3-based techniques should be introduced, as they have shown their merit in design and assessment. This includes the possibility of custom expert-defined rules, as the rules are currently embedded. The rules could also suggest possible relationships between potential evidence. Beyond Z3, the architecture allows for analyses based on other techniques in dedicated modules (e.g., a module utilising process mining). Finally, the tool's usability should be evaluated with domain experts.

Acknowledgement. The work was supported by the Grant Agency of Masaryk University (GAMU) project "Forensic Support for Building Trust in Smart Software Ecosystems", registration number MUNI/G/1142/2022. It was also co-founded by the European Union under Grant Agreement No. 101087529. Views and opinions expressed are however those of the author(s) only and do not necessarily reflect those of the European Union or European Research Executive Agency. Neither the European Union nor the granting authority can be held responsible for them.

References

1. Bjørner, N., de Moura, L., Nachmanson, L., Wintersteiger, C.M.: Programming Z3, pp. 148–201. Springer, Cham (2019)
2. Casey, E., Nikkel, B.: Forensic Analysis as Iterative Learning. In: Keupp, M. (ed.) The Security of Critical Infrastructures, pp. 177–192. Springer, Cham (2020). https://doi.org/10.1007/978-3-030-41826-7_11
3. CESG: Good Practice Guide No. 18: Forensic Readiness. Guideline, National Technical Authority for Information Assurance, United Kingdom (2015)
4. Daubner, L., Macak, M., Matulevičius, R., Buhnova, B., Maksović, S., Pitner, T.: Addressing insider attacks via forensic-ready risk management. J. Inf. Secur. Appl. **73**, 103433 (2023)
5. Daubner, L., Matulevičius, R., Buhnova, B.: A model of qualitative factors in forensic-ready software systems. In: Nurcan, S., Opdahl, A.L., Mouratidis, H., Tsohou, A. (eds.) RCIS 2023, pp. 308–324. Springer, Cham (2023). https://doi.org/10.1007/978-3-031-33080-3_19
6. Daubner, L., Matulevičius, R., Buhnova, B., Pitner, T.: BPMN4FRSS: an BPMN extension to support risk-based development of forensic-ready software systems. In: Kaindl, H., Mannion, M., Maciaszek, L.A. (eds.) ENASE 2022. CCIS, vol. 1829, pp. 20–43. Springer, Cham (2023). https://doi.org/10.1007/978-3-031-36597-3_2
7. Daubner, L., Matulevičius, R.: Risk-oriented design approach for forensic-ready software systems. In: The 16th International Conference on Availability, Reliability and Security. ACM (2021)
8. Dzurenda, P., et al.: Privacy-preserving solution for vehicle parking services complying with EU legislation. PeerJ Comput. Sci. **8**, e1165 (2022)
9. Erol-Kantarci, M., Mouftah, H.T.: Smart grid forensic science: applications, challenges, and open issues. IEEE Commun. Mag. **51**(1), 68–74 (2013)
10. Grispos, G., Glisson, W.B., Choo, K.K.R.: Medical cyber-physical systems development: a forensics-driven approach. In: IEEE/ACM International Conference on Connected Health: Applications, Systems and Engineering Technologies, pp. 108–113 (2017)
11. Jürjens, J.: Model-based security engineering with UML. In: Aldini, A., Gorrieri, R., Martinelli, F. (eds.) FOSAD 2005 2004. LNCS, vol. 3655, pp. 42–77. Springer, Heidelberg (2005). https://doi.org/10.1007/11554578_2
12. Maksović, S.: Model-based analysis of forensic-ready software systems. Bachelor's thesis, Masaryk University (2023). https://is.muni.cz/th/w43li/
13. Matulevičius, R.: Fundamentals of Secure System Modelling. Springer, Cham (2017). https://doi.org/10.1007/978-3-319-61717-6
14. McKemmish, R.: When is digital evidence forensically sound? In: Ray, I., Shenoi, S. (eds.) Advances in Digital Forensics IV, pp. 3–15. Springer, Boston (2008). https://doi.org/10.1007/978-0-387-84927-0_1
15. Moura, L.D., Bjørner, N.: Z3: an efficient SMT solver. In: Proceedings of the Theory and Practice of Software, 14th International Conference on Tools and Algorithms for the Construction and Analysis of Systems, pp. 337–340 (2008)
16. Pasquale, L., Alrajeh, D., Peersman, C., Tun, T., Nuseibeh, B., Rashid, A.: Towards forensic-ready software systems. In: Proceedings of the 40th International Conference on Software Engineering: NIER, pp. 9–12. ACM (2018)
17. Pasquale, L., Spoletini, P., Salehie, M., Cavallaro, L., Nuseibeh, B.: Automating trade-off analysis of security requirements. Requirements Eng. **21**(4), 481–504 (2016)

18. Pullonen, P., Matulevičius, R., Bogdanov, D.: PE-BPMN: privacy-enhanced business process model and notation. In: Carmona, J., Engels, G., Kumar, A. (eds.) BPM 2017. LNCS, vol. 10445, pp. 40–56. Springer, Cham (2017). https://doi.org/10.1007/978-3-319-65000-5_3
19. Rivera-Ortiz, F., Pasquale, L.: Automated modelling of security incidents to represent logging requirements in software systems. In: Proceedings of the 15th International Conference on Availability, Reliability and Security. ACM (2020)
20. Rowlingson, R.: A ten step process for forensic readiness. Int. J. Digit. Evid. **2**, 1–28 (2004)
21. Sedláček, T.: Web-based editor for BPMN4FRSS models. Bachelor's thesis, Masaryk University (2023). https://is.muni.cz/th/oiby0/
22. Simou, S., Kalloniatis, C., Gritzalis, S., Katos, V.: A framework for designing cloud forensic-enabled services (CFES). Requirements Eng. **24**(3), 403–430 (2019)
23. Tan, J.: Forensic readiness. Technical report, @stake, Inc. (2001)
24. Toots, A., et al.: Business process privacy analysis in pleak. In: Hähnle, R., van der Aalst, W. (eds.) FASE 2019. LNCS, vol. 11424, pp. 306–312. Springer, Cham (2019). https://doi.org/10.1007/978-3-030-16722-6_18
25. Yin, Y., Tateiwa, Y., Wang, Y., Katayama, Y., Takahashi, N.: Inconsistency analysis of time-based security policy and firewall policy. In: Duan, Z., Ong, L. (eds.) ICFEM 2017. LNCS, vol. 10610, pp. 447–463. Springer, Cham (2017). https://doi.org/10.1007/978-3-319-68690-5_27

Control and Monitoring of Software Robots: What Can Academia and Industry Learn from Each Other?

Kelly Kurowski[1], Antonio Martínez-Rojas[2(✉)], and Hajo A. Reijers[1]

[1] Department of Information and Computing Sciences, Utrecht University, Utrecht, The Netherlands
k.kurowski@students.uu.nl, h.a.reijers@uu.nl

[2] Department of Computer Languages and Systems, University of Seville, Seville, Spain
amrojas@us.es

Abstract. Robotic Process Automation (RPA) has witnessed significant growth, becoming widely adopted in practice. This surge in the use of RPA technology has given rise to new challenges, particularly concerning the effective control and monitoring of software robots. Ideally, academia and industry would work together on developing new RPA capabilities, but both domains operate rather separately. In this paper, we employ an explorative approach to examine how academic theories can improve industrial RPA practices and vice versa. By analyzing both academic literature and leading RPA platforms, we present four recommendations for academia, four for industry, and a general recommendation aiming to advance the collaboration between them.

Keywords: Robotic Process Automation · Monitoring · Maintenance · Systematic Literature Review · Industry Review · Challenges

1 Introduction

Robotic Process Automation (RPA) uses software agents to automate tasks by emulating human interactions within digital systems, reducing costs, and minimizing errors [19]. Its adoption across various sectors has increased, but challenges, particularly in robot control and monitoring, necessitate further exploration [19]. Challenge 15 in [19] highlights the issue of effectively managing RPA robots, emphasizing the necessity for improved monitoring techniques to ensure continued performance and accuracy. Current academic and commercial solutions, such as UiPath Orchestrator [3] and BluePrism Control Room [2], reveal significant gaps in addressing these challenges.

This research was supported by the EQUAVEL project PID2022-137646OB-C31 funded by MICIU/AEI/10.13039/501100011033 and by FEDER, UE; the grant FPU20/05984 funded by MICIU/AEI/10.13039/501100011033 and by FSE+, and its mobility grant EST23/00732.

J. Araújo et al. (Eds.): RCIS 2024, LNBIP 514, pp. 56–64, 2024.
https://doi.org/10.1007/978-3-031-59468-7_7

Table 1. Classification framework for the related work.

Study	RPA Focus	Topic Specialization	Academic Approaches	Industrial Platforms	Identify Challenges	Industry vs. Academia
[15]	✓	Sustainability	✓	X	X	X
[14]	✓	General	✓	X	X	X
[12]	✓	Intelligent Document Processing	✓	✓	✓	X
[20]	✓	General	✓	X	✓	X
[6]	✓	Socio-Human Implications	✓	X	✓	X
[4]	✓	Digital Transformation	✓	X	✓	X
[18]	✓	Use of RPA in covid-19	✓	X	X	X
[17]	✓	Document Classification	✓	X	X	X
[16]	✓	Industry 4.0	X	✓	X	X
[5]	✓	General	✓	✓	✓	X
[9]	✓	General	✓	X	X	X
Our study	✓	RPA Control + Monitoring	✓	✓	✓	✓

In response to these gaps, this paper conducts a comprehensive review targeting the control and monitoring aspects of RPA, aiming to foster better collaboration between academia and industry. By systematically examining existing academic proposals and RPA platform capabilities, we propose nine recommendations to enhance the alignment of research with practical RPA application needs.

This paper is structured as follows: Related works are discussed in Sect. 2, the research methodology in Sect. 3, analysis of findings in Sect. 4, recommendations in Sect. 5, and conclusion and final remarks in Sect. 6.

2 Related Work

In this section, a systematic scrutiny of the literature review pertinent to RPA is conducted. Utilizing the *SCOPUS* database, specifically within the *Computer Science* subject area and focusing on *article titles*. The search query utilized was: *("literature review" OR "systematic mapping study") AND (rpa OR "robotic process automation")*. This search resulted in 11 review studies, presented in Table 1. This classification framework encapsulates critical dimensions of each reviewed study, encompassing: alignment with RPA, topic specialization, focus on academic approaches, examination of industrial platforms, identification of existing challenges or gaps within the field, and comparative analysis between industry and academia.

The synthesized findings reveal a diverse range of focuses amongst the reviews, with a significant portion delving into specific RPA-related topics like sustainability and industry 4.0, predominantly within the academic sphere. A notable deficiency is observed in the comparative analyses between industrial and academic perspectives, highlighting a gap in the current literature. The least explored area of RPA, robot control and monitoring is explored in this paper. While linking academic research with industrial practices, this review fosters synergies between the two spheres and advances knowledge with practical ideas and recommendations beneficial to both sectors.

3 Research Method

3.1 Literature Review

In this study, our primary objective was to delve into the control and monitoring phase of RPA within an academic context. We executed a Systematic Literature Review (SLR) following the methodology proposed by [11].

The *Scopus* search engine served as our primary tool, setting six inclusion and exclusion criteria. (1) We scrutinized article titles, abstracts, and keywords with mandatory and optional terms. (2) As there was no relevant work in RPA before 2019, the search spanned from 2019 to 2023. (3) Studies must be written in English, (4) published as journals or conference papers, and (6) they must propose an approach to control and monitoring in RPA.

Our search methodology utilized three targeted queries. Initiating with the query (**"Robotic Process Automation" AND ("Monitoring" OR "Maintenance"**)), we identified 79 papers. From these, we selected [7,13] based on their alignment with criterion 6, pertaining to the control and monitoring in RPA, after examining the relevance of the top 40 articles. Enhancing our search to encompass "Orchestration" with the query (**"Robotic Process Automation" AND ("Maintenance" OR "Orchestration" OR "Monitoring"**)), we retrieved 72 documents. From the relevant entries, we selected [8,21], applying the same relevance criteria. Further refining our search, the query (**"Robotic Process Automation" AND ("Governance" OR "Orchestration" OR "Monitoring" OR "Maintenance"**)) led to 106 documents. After reviewing the top 40 for their pertinence, particularly against criterion 6, we chose [10].

Altogether, this methodical search strategy resulted in five high-quality, recent studies that meet our specified research parameters.

3.2 Industrial Review

To assess the current state of industrial landscape, we selected RPA vendors from the 2023 Gartner Magic Quadrant[1] and Forrester Wave[2] reports. We focused on

[1] Gartner Magic Quadrant RPA: www.uipath.com/es/resources/automation-analyst-reports/gartner-magic-quadrant-robotic-process-automation.

[2] Forrester Wave RPA: https://www.uipath.com/resources/automation-analyst-reports/forrester-wave-rpa.

the intersection of vendors classified as *Leaders* in both reports, leading us to *SS&C Blue Prism* [2], *Automation Anywhere* [1] (AA), and *UiPath* [3]. Each of the RPA vendors provides documentation for their platforms, which became the focus of our research.

4 Analysis

To analyze academic and industrial approaches, we focused on the following features: (1) *Monitoring*: Real-time tracking of system performance and health. (2) *Scheduling*: Ability to schedule robot jobs. (3) *Exception Handling*: Management of unexpected errors or exceptional conditions. (4) *Logging*: Recording of events, actions, or messages during process execution. (5) *Information Display*: Presentation of relevant information to users or developers. Resulting in a concise set for cross-platform and cross-study comparison.

4.1 Academic Approaches

Five academic studies have been analyzed[3], classified into literature reviews, case studies, and technical proposals. These include Hartikainen et al.'s review of robot maintenance [7], Kedziora et al.'s governance case study within a banking context [10], Šimek et al.'s exploration of Robotic Service Orchestration in HR [21], Martínez-Rojas et al.'s development of AIRPA for enhanced lifecycle management [13], and Hwang et al.'s MIORPA for open-source robot control [8].

Key findings reveal a prevalent manual approach across academic works concerning scheduling, monitoring, exception handling, and logging, indicating a gap in automation and efficiency. For instance, manual scheduling and monitoring are common despite the generation of extensive data, and exception handling largely relies on human intervention. Moreover, while academic approaches suggest advanced logging and information display mechanisms, they rarely integrate with existing RPA platforms.

4.2 RPA Platforms

The examination of UiPath, SS&C Blue Prism, and Automation Anywhere (AA) platforms[4] highlighted differences in scheduling, monitoring, exception handling, logging, and information display. UiPath stands out for its dynamic workload distribution and real-time metrics, whereas Blue Prism and AA offer robust scheduling and error analysis capabilities.

Comparatively, all platforms provide comprehensive logging and monitoring features but differ in their approach to exception handling and information display. The analysis suggests that while AA and Blue Prism share similarities, UiPath excels in offering more detailed insights and performance indicators.

[3] Full academic analysis described in: https://doi.org/10.5281/zenodo.10818625.

[4] Full industrial analysis described in: https://doi.org/10.5281/zenodo.10818625.

5 Discussion

This discussion delves into actionable recommendations for both industry and academia, acknowledging the symbiotic relationship between them (cf. Fig. 1).

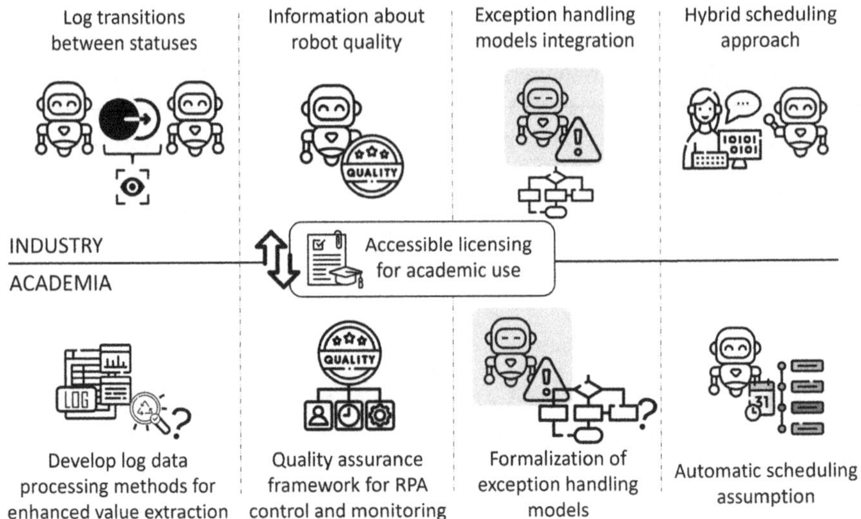

Fig. 1. Recommendations for industry and academia. The ones for industry are shown at the top and for academia at the bottom. They are vertically related by the topics of logging, information display, exception handling, and scheduling, in that order.

5.1 Recommendations for Industry

To enhance the capabilities of existing RPA platforms, several key recommendations emerge:

Log Transitions Between Statuses: It is noted that similar data, such as status, start time, or end time, is stored by the platforms. A valuable avenue for enhancing robot monitoring lies in adopting the academic proposal to capture evidence of transitions between robot statuses [13]. This method has the potential to (1) pinpoint areas in need of reinforcement, (2) enhance the detection of the causes of exceptions, and (3) provide clarity regarding the performance of each step executed by the robot.

Information About Robot Quality: RPA platforms mainly focus on presenting information or forecasts on the digital workforce utilization or its performance in dashboards. Nevertheless, the idea of *quality assessment* is a worthwhile consideration [7]. By incorporating quality indicators for robot jobs on platform dashboards, industrial RPA platforms become more suitable to proactively monitor robots, i.e., to enable the early detection of maintenance requirements, particularly for robots based on AI models.

Exception Handling Models Integration: Highlighting exceptions is standard practice in the industry. However, it should be noted that the incorporation of exception handling models in one of the academic studies [21], demonstrated high effectiveness in reducing exceptions during testing. If RPA vendors were to integrate similar models into their platforms, then this may contribute to smoother and more reliable automation processes.

Hybrid Scheduling Approach: RPA platforms focus on automatic scheduling solutions. However, the academy proposes manual methods, most notably the one based on specific business rules [8]. By incorporating hybrid scheduling approaches, which integrate business rule-based prioritization with automated scheduling, industrial RPA platforms may be enhanced. Such an approach allows for the automation of routine scheduling while involving business experts for more complex scenarios; this offers a balanced and efficient scheduling strategy.

5.2 Recommendations for Academia

Academic contributions can be essential in shaping the future of RPA, here are the key recommendations derived from our analysis:

Develop Log Data Processing Methods: Major industrial platforms are generating substantial amounts of data, including *status updates*, *robot execution ID*, and more. This wealth of information presents a significant opportunity for academia to develop methods for extracting value. For example, methods to identify similarities within execution logs would be useful, so that it becomes feasible to modularize common components and increase reusability in RPA processes.

Quality Assurance Framework for RPA: Our analysis of industrial platforms highlights the lack of quality metrics for robot work. Although the work of [7] emphasized the importance of quality assessment, it is the only one to mention it. Academia can play a crucial role in developing a comprehensive quality assurance framework for RPA. By addressing both the effectiveness of robots' tasks and the quality of the AI models in use, a quality framework could provide organizations with a systematic approach to evaluate and improve their overall RPA solutions.

Formalization of Exception Handling Models: Exception management models are a novelty proposed by the academy However, academics are encouraged to deepen the formalization of exception handling models for RPA processes. The definition of systematic methods to create these models remains an unexplored area. This gap represents an opportunity for academia to generate new knowledge, while industry can benefit from the operationalization and implementation of standardized exception management practices, thus improving the reliability of RPA processes.

Assumption of Automatic Scheduling: Building on the insights presented in this paper, it seems wise that academics build on the premise that the industry leans toward extensively automating scheduling processes. Consequently, the recommendation is to refrain from suggesting manual methods, although the industrial, automated approach might not be universally applicable to all cases. An opportunity here for academia is to explore hybrid scheduling approaches. This involves automated scheduling validated by business experts, combining automation and human expertise. Such a hybrid approach would not only address the need for automation in routine scheduling but would also integrate nuanced decision-making capabilities of business experts for scenarios that require specific criteria.

In the center of Fig. 1, we also positioned what we believe to be the "passage" for a more fruitful collaboration between academia and industry, i.e. *accessible licensing for academic use.* It should be noted that 2/3 of the main RPA platforms require a pay-to-use method (cf. Table 1). A less restrictive licensing structure of RPA platforms would empower researchers to directly implement their proposals on these platforms, fostering the creation of add-ons, plug-ins, and enhancements. Such a collaborative process would not only help to refine academic methodologies but also clarify how the capabilities of industrial platforms can be extended.

6 Conclusion

To address a clear and pressing need for enhancing the control and monitoring capabilities of RPA technology, we followed an exploratory approach to juxtapose academic proposals in this area with the current capabilities that RPA platforms offer. This led to the formulation of nine recommendations. These are thought to be helpful for RPA vendors to pick up on promising academic proposals. Additionally, they can guide academic researchers to tune their work to industrial practice.

We wish to acknowledge a number of limitations to our work. First of all, our reliance on platform documentation, rather than direct access to the platforms, limits the accuracy of our assessment. Academic access to RPA platforms, cf. our "passage" recommendation, could facilitate a more comprehensive and in-depth analysis. Secondly, the sample size of RPA vendors is small, which can negatively affect the findings' generalizability, prompting consideration for a larger-scale study. Finally, we incorporated a single search engine for our SLR, which can have a negative impact on the diversity and robustness of our findings.

We hope that this work contributes to a closer collaboration between academics working on RPA topics and vendors of RPA technology. While our approach is exploratory in trying to reconcile efforts on both sides, it may also inspire the development of more structural modes of interaction and knowledge exchange.

References

1. Automation Anywhere Control Room Documentation. https://docs.automationanywhere.com/bundle/enterprise-v11.3/page/enterprise/topics/control-room/getting-started/using-control-room.html
2. BluePrism Control Room: Hub 4.6 and Control Room 4.6 user guide. https://bpdocs.blueprism.com/hub-interact/4-6/en-us/home-control-room.htm
3. UiPath Orchestrator Documentation. https://docs.uipath.com/orchestrator/standalone/2023.4/user-guide/introduction
4. Afriliana, N., Ramadhan, A.: The trends and roles of robotic process automation technology in digital transformation: a literature. J. Syst. Manag. Sci. **12**(3), 51–73 (2022)
5. Enríquez, J.G., Jiménez-Ramírez, A., Domínguez-Mayo, F.J., García-García, J.A.: Robotic process automation: a scientific and industrial systematic mapping study. IEEE Access **8**, 39113–39129 (2020)
6. Harmoko, H., Ramírez, A.J., Enríquez, J.G., Axmann, B.: Identifying the socio-human inputs and implications in robotic process automation (RPA): a systematic mapping study. In: Marrella, A., et al. (eds.) BPM 2022. LNBIP, vol. 459, pp. 185–199. Springer, Cham (2022). https://doi.org/10.1007/978-3-031-16168-1_12
7. Hartikainen, E., Hotti, V., Tukiainen, M.: Improving software robot maintenance in large-scale environments-is center of excellence a solution? IEEE Access **10**, 96760–96773 (2022). https://doi.org/10.1109/ACCESS.2022.3205420
8. Hwang, M.H., et al.: MIORPA: middleware system for open-source robotic process automation. J. Comput. Sci. Eng. **14**(1), 19–25 (2020)
9. Ivančić, L., Suša Vugec, D., Bosilj Vukšić, V.: Robotic process automation: systematic literature review. In: Di Ciccio, C., et al. (eds.) BPM 2019. LNBIP, vol. 361, pp. 280–295. Springer, Cham (2019). https://doi.org/10.1007/978-3-030-30429-4_19
10. Kedziora, D., Penttinen, E.: Governance models for robotic process automation: the case of Nordea bank. J. Inf. Technol. Teach. Cases **11**(1), 20–29 (2021). https://doi.org/10.1177/2043886920937022
11. Kitchenham, B., Brereton, O.P., Budgen, D., Turner, M., Bailey, J., Linkman, S.: Systematic literature reviews in software engineering-a systematic literature review. Inf. Softw. Technol. **51**(1), 7–15 (2009)
12. Martínez-Rojas, A., López-Carnicer, J., González-Enríquez, J., Jiménez-Ramírez, A., Sánchez-Oliva, J.: Intelligent document processing in end-to-end RPA contexts: a systematic literature review. In: Bhattacharyya, S., Banerjee, J.S., De, D. (eds.) Confluence of Artificial Intelligence and Robotic Process Automation, vol. 335, pp. 95–131. Springer, Singapore (2023). https://doi.org/10.1007/978-981-19-8296-5_5
13. Martínez-Rojas, A., Sánchez-Oliva, J., López-Carnicer, J.M., Jiménez-Ramírez, A.: AIRPA: an architecture to support the execution and maintenance of AI-powered RPA robots. In: González Enríquez, J., Debois, S., Fettke, P., Plebani, P., van de Weerd, I., Weber, I. (eds.) BPM 2021. LNBIP, vol. 428, pp. 38–48. Springer, Cham (2021). https://doi.org/10.1007/978-3-030-85867-4_4
14. Moreira, S., Mamede, H.S., Santos, A.: Process automation using RPA-a literature review. Procedia Comput. Sci. **219**, 244–254 (2023)
15. Patrício, L., et al.: Literature review of decision models for the sustainable implementation of robotic process automation. Procedia Comput. Sci. **219** (2023)
16. Ribeiro, J., Lima, R., Eckhardt, T., Paiva, S.: Robotic process automation and artificial intelligence in industry 4.0-a literature review. Procedia Comput. Sci. **181**, 51–58 (2021)

17. Ribeiro, J., Lima, R., Paiva, S.: Document classification in robotic process automation using artificial intelligence—a preliminary literature review. In: Sharma, H., Gupta, M.K., Tomar, G.S., Lipo, W. (eds.) Communication and Intelligent Systems. LNNS, vol. 204, pp. 211–221. Springer, Singapore (2021). https://doi.org/10.1007/978-981-16-1089-9_18
18. Štorek, F., Basl, J., Doucek, P.: Use of robotic process automation tools during and after covid-19 pandemic from the industry perspective: literature review. IDIMT **2021**, 55–60 (2021)
19. Syed, R., et al.: Robotic process automation: contemporary themes and challenges. Comput. Ind. **115**, 103162 (2020)
20. Wewerka, J., Reichert, M.: Robotic process automation - a systematic mapping study and classification framework. Enterprise Inf. Syst. **17**, 1–38 (2021). https://doi.org/10.1080/17517575.2021.1986862
21. Šimek, D., Šperka, R.: How robot/human orchestration can help in an HR department: a case study from a pilot implementation. Organizacija **52**(3), 204–217 (2019). https://doi.org/10.2478/orga-2019-0013

Creating a Web-Based Viewer for an ADOxx-Based Modeling Toolkit

Ilia Bider[1,2(✉)] and Siim Langel[2]

[1] DSV, Stockholm University, Postbox 7003, 16407 Kista, Sweden
ilia@dsv.su.se

[2] ICS,University of Tartu, Narva Mnt 18, 51009 Tartu, Estonia
ilia.bider@ut.ee, siimlangel11@gmail.com

Abstract. ADOxx is an environment that allows the creation of a toolkit for a specific modeling technique while using limited resources for development. The result is a toolkit for professional modelers that can be run on Windows, Linux, or MAC. However, such a toolkit is not very friendly for non-IT related stakeholders, as it is primarily aimed at developing the models; it also requires non-trivial installation. This paper is devoted to the project of designing a WEB-based viewer that allows a stakeholder to view a package of models created in an ADOxx-based toolkit. The viewer discussed in this paper was developed for a specific modeling technique called Fractal Enterprise Model (FEM). However, the discussion is of interest not only for modelers using FEM but also for the developers of other ADOxx-based tools. The paper discusses the structure and functionality of the FEM viewer, which can be reused for other toolkits. The authors aim to demonstrate the FEM viewer during the conference.

Keywords: Modeling tools demonstration · ADOxx · model viewer · Fractal Enterprise Model · FEM

1 Motivation

The tool - FEM viewer - discussed in this paper was developed for a specific modeling language and notation called the Fractal Enterprise Model (FEM) [1]. However, the need for a new tool was dictated by some drawbacks of the modeling tool, called the FEM toolkit [2], which was created using the ADOxx [3] meta-modeling environment. Therefore, the functionality and structure of the FEM viewer might be of interest not only to professionals using or aiming at using FEM, but also to the developers of toolkits for other modeling languages who use the ADOxx environment. As the source code of the FEM viewer is available under an open-source license, they could even use some parts of the developed source code when creating a viewer as a companion for their modeling tools developed using ADOxx.

A tool produced using ADOxx is a tool for modelers, in the first hand. It is not always easy to install the tool, especially on non-Windows machines, and using it requires some skills, even when just browsing a set of already existing models. Thus, giving this tool

J. Araújo et al. (Eds.): RCIS 2024, LNBIP 514, pp. 65–73, 2024.
https://doi.org/10.1007/978-3-031-59468-7_8

to the non-modeling stakeholders does not work for everyone. Additional difficulties can be experienced in a corporate environment where installing new software is strictly regulated. An alternative method of providing stakeholders with a package of models as printed PDF sheets has its own drawbacks. While you can easily investigate an individual model, the possibility of going from one model to another using references is lost. The PDF prints also do not show all properties assigned to the elements of the model.

The FEM viewer was developed to solve the problem of giving access to the models to a non-technical user. It is a web-based tool that offers the possibility for its users to browse through a set of interconnected FEMs. The package of models can be exported from the FEM toolkit and imported into the FEM viewer. After that, a user or group of users can be given access to the package. The user can easily browse through the set of models following the links connecting the models.

The rest of the paper is structured according to the following plan. In Sect. 2, we give the minimum information on FEM. In Sect. 3, we describe the functionality of the FEM toolkit. In Sect. 4, we describe the technical implementation of it. Section 5 has some historical information and plans for the future.

2 Fractal Enterprise Model

In this section, we present a minimal description of the Fractal Enterprise Model (FEM), just enough for reading and understanding this paper. Readers interested in knowing more are referred to [1, 2, 4]. FEM is a modeling language and notation for representing the operational activities of an organization. Examples of FEM diagrams are represented in Figs. 1 and 2. FEM has four main concepts and several types of relations between them. The concepts and their representations are as follows:

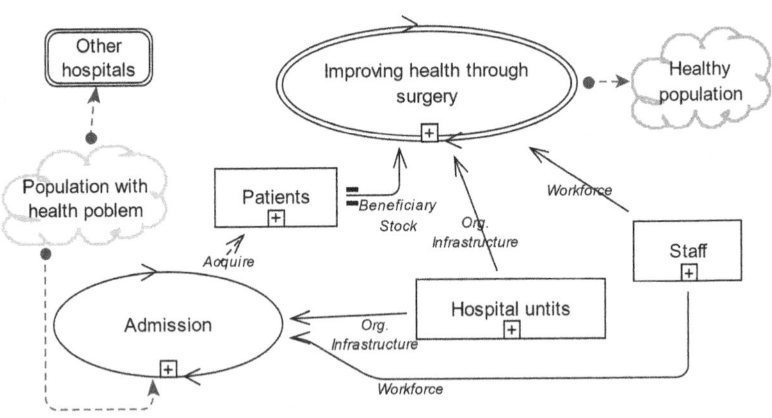

Fig. 1. Model of a hospital – an overview

1. A process – repetitive behavior – is represented as an oval. The double line border means this is a primary process that produces value for an external beneficiary.

2. An asset – a set of things or actors that are used in a process or can be managed by a process – is represented as a rectangle. A rectangle with dash-doted borders represents a tacit asset (that is inside the head of some actors).
3. An external pool – a set of things or actors that can be used for adding elements to own assets – is represented as a cloud shape. Also, elements can be added to an external pool by a process.
4. An external actor – an individual or a set of actors that can draw elements from a pool or add elements to a pool – is represented as a rectangle with rounded corners. The double border line is used if there is more than one external actor. The concept is used to represent competitors or collaborators.

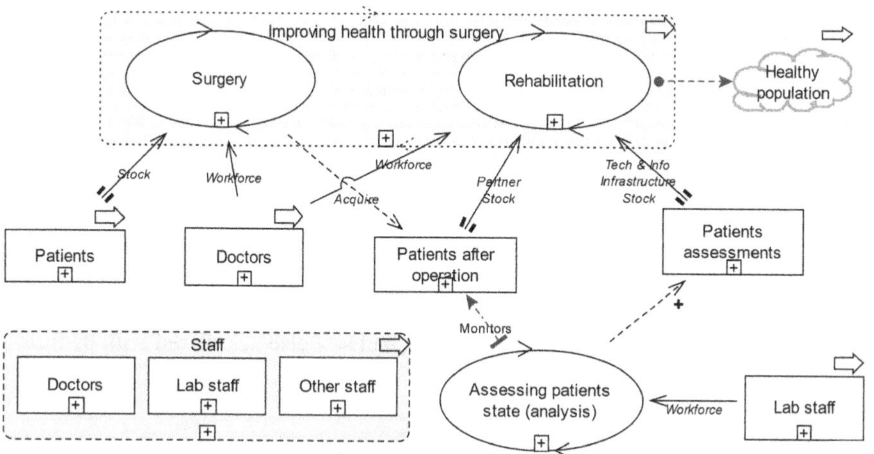

Fig. 2. Decomposition and details of hospital activities

Besides the shapes used for representing the concepts above, the FEM toolkit uses a *Note* shape to add comments to the diagrams (not shown in the Fig. 1 and 2). Each shape in a diagram has properties; some of them are visualized in the form of the shape, and others are invisible until the user explicitly asks to show them. The FEM toolkit has several properties that allow it to connect several diagrams and highlight some essential elements in them. Connecting is done using the concept of ghost; a ghost represents the second, third, etc., appearance of the same element in the same or different diagram. The ghost shape has a thick arrow in its upper right corner; clicking on it will move the focus to the first occurrence of the element, which might be in another diagram. There is also a possibility of finding all occurrences of a specific element.

Highlighting is done either through a specially designated background color or a specially designated border color (see cloud shapes in Fig. 1). The designation of colors is done in special kinds of diagrams, called FEM subclassing and Border sub-classing, respectively [2]. The FEM toolkit allows to navigate from the shape to the FEM subclaass or border subclass to which it belongs. It also allows finding all shapes that belong to a specific subclass.

The FEM toolkit also allows the decomposition/specialization of an element by making it a group and putting other shapes inside the group shape. The group gets a shape as a rectangle with a dashed (specialization of processes or decomposition of other elements) or dotted (decomposition of processes) border; see Fig. 2.

3 Functionality of FEM Viewer

Most of the functionality described in this section is general; it can be used in a viewer designed for a different ADOxx language. The exception is the functionality related to navigation between the models (see Sect. 3.2), which is specific to the FEM toolkit.

3.1 User Management

The FEM viewer has **three types of users**: (1) *Administrator*, (2) *Developer* and (3) *Viewer*. An administrator and developer can **manage users**, while this functionality is inaccessible to viewers. An administrator can create new users of any type, and a developer can create users of the type *Viewer*. Also, both can delete users that they have created.

An administrator and developer can **export** a package of models from the FEM toolkit and **import** it to the FEM viewer. A viewer cannot use the import functionality. An administrator and developer can **share** an imported package with any user they have created. They can also cancel sharing the package. All users can **view** a package of models they have imported or a package that somebody else has shared with them.

3.2 Viewing Models

Viewing a package of models is the primary operation of the FEM viewer. All types of users can view models they have imported or which have been shared with them. The viewing mode is shown in Fig. 3. The viewer window is divided into three parts. The upper part shows: (1) the name of the package (Model tree), (2) the name of the model under investigation and its type in parentheses, and (3) a slider that allows to change the size of the model, and a button that allows the user to move to their dashboard.

The lower left part lists all the models in the package and allows the user to choose one for viewing. The name of the viewed model is underlined, and the box to the left is checked. The lower right part presents the model the user selected for viewing. Figure 3 has the model presented in Fig. 1 as chosen for viewing.

Clicking on a shape in the model will highlight the respective element and bring a new window that shows additional information about the element. Figure 4 shows the window for the pool *Healthy population* in the *Overview* model. The window displays the properties of the element. Some properties can be seen visually in the model, like the class of the element and short description; others are hidden, e.g., long description. Which properties to show is steered by a special configuration file. Besides properties, the pop-up window includes navigational mechanisms. *All occurrences* shows other models in the package that include the same element. Clicking on the model's name in

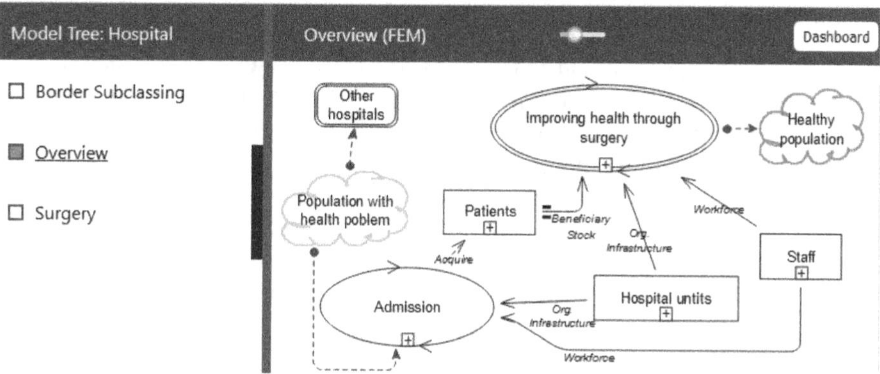

Fig. 3. Hospital overview model in the FEM viewer

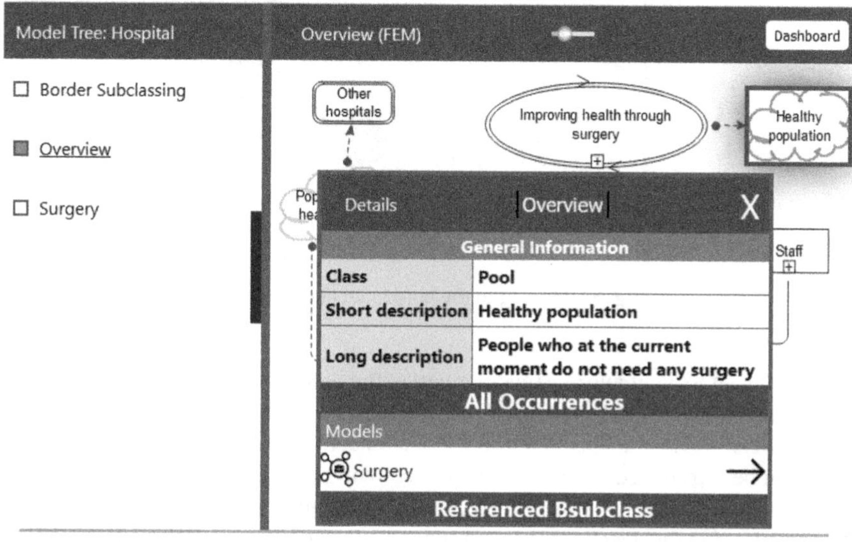

Fig. 4. A popup window with properties and references

the *All Occurrences* submenu will bring forward the model where the *Healthy population* elements are highlighted.

Clicking on *References Bsubclass* will bring the *Border Subclassing* model to the viewer and highlight the relevant border subclass, *Output pool* in our case that is marked by the green border color in Fig. 5. At the same time, a pop window appears that gives the user the possibility to find all elements that are related to the *Output Pull* border subclass (see Fig. 5).

3.3 Importing and Sharing the Models

Users of type *Administrator* and *Developer* can import a package of models exported from the FEM toolkit. To import a package of models, an XML file export should be completed in the FEM toolkit. This is a standard function provided by ADOxx, and it is available in all toolkits developed with this metamodeling environment. In principle, the XML export file contains all information about the models included in the package. However, reproducing the graphical information is not a trivial task; therefore, we used another approach for importing graphical representation into the FEM viewer. Namely, we produce and import SVG images of all models included in the package.

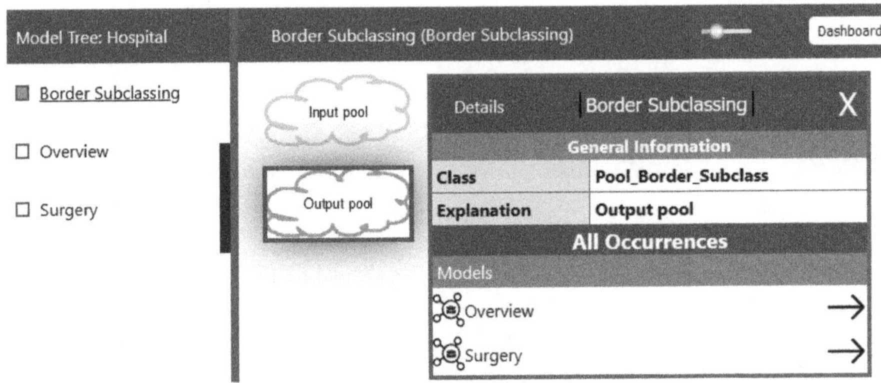

Fig. 5. Moving to the border subclass model

Producing an SVG image of a model is part of the functionality available via ADOxx. To make importing easier, we created a function that produces SVG images for a set of chosen models in one go. This function is not connected to the particulars of FEM, and its code can be introduced in any toolkit created with the help of ADOxx.

Importing a package is done via the screen presented in Fig. 6 that appears when pressing *Upload* in the dashboard of an *Administrator* or *Developer.* Pressing the button *Upload* underneath the info on which files should be included will create a new modeling package in the FEM viewer. Note that the *Dashboard* for this type of user contains the list of all imported packages and packages shared with the user by somebody else. It also includes all users they created. An imported package can be shared with any user that appears in the dashboard. For the viewer, the dashboard contains only the list of packages their developer or administrator has shared with them.

From the list of imported or shared packages, a user can pick a package and open it in the viewer to access the functionality presented in Figs. 3, 4, and 5.

Note that SVG images are used to present the imported models in the viewer. Information from the XML file is used to show model elements' properties and arrange navigation between the models through references. Information on the properties and references is extracted from the imported XML.

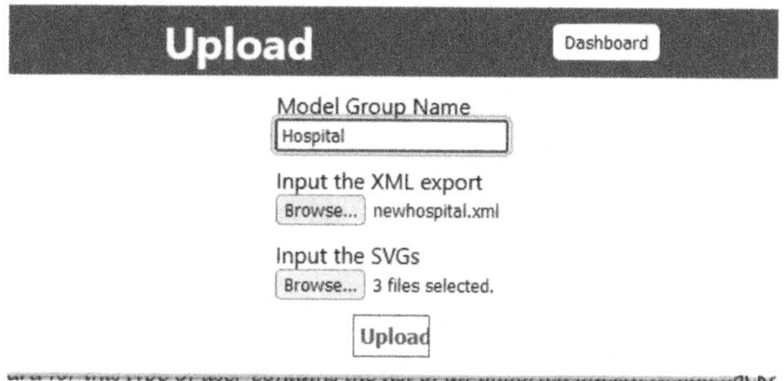

Fig. 6. Upload/import function

4 Technical Implementation

The FEM viewer is a WEB-based application developed using javascript tools. It runs on the Linux server and users MySQL database to manage users and their connec-tion to model packages that they have imported or that have been shared with them by their developer. The way the user communicates with the client software on a browser and the client communicates with the server is presented in Fig. 7. The com-plete source code of the viewer is available on GitHub [5] under an open-source li-cense.

The following tools were used for implementation of FEM viewer:

1. React.js [6] was used for building user interface on the client side (WEB browser). React is a declarative JavaScript library.
2. Node.js [7] was used for developing the server side of the FEM viewer.
3. Express.js [8], which is a minimalistic framework built upon Node.js, was used to develop the major functionality on the server side, such as uploading files for im-port or user authentication.
4. Prisma.js [9], which is a Node.js library, was used to handle persistent data related to users. It provides objects mapping to their representations in a relational data-base, thus removing the needs for writing SQL code.
5. Pasport.js [10], which is a Node.js library, was used to provide session authentica-tion for the users. It can work with any Express.js application.

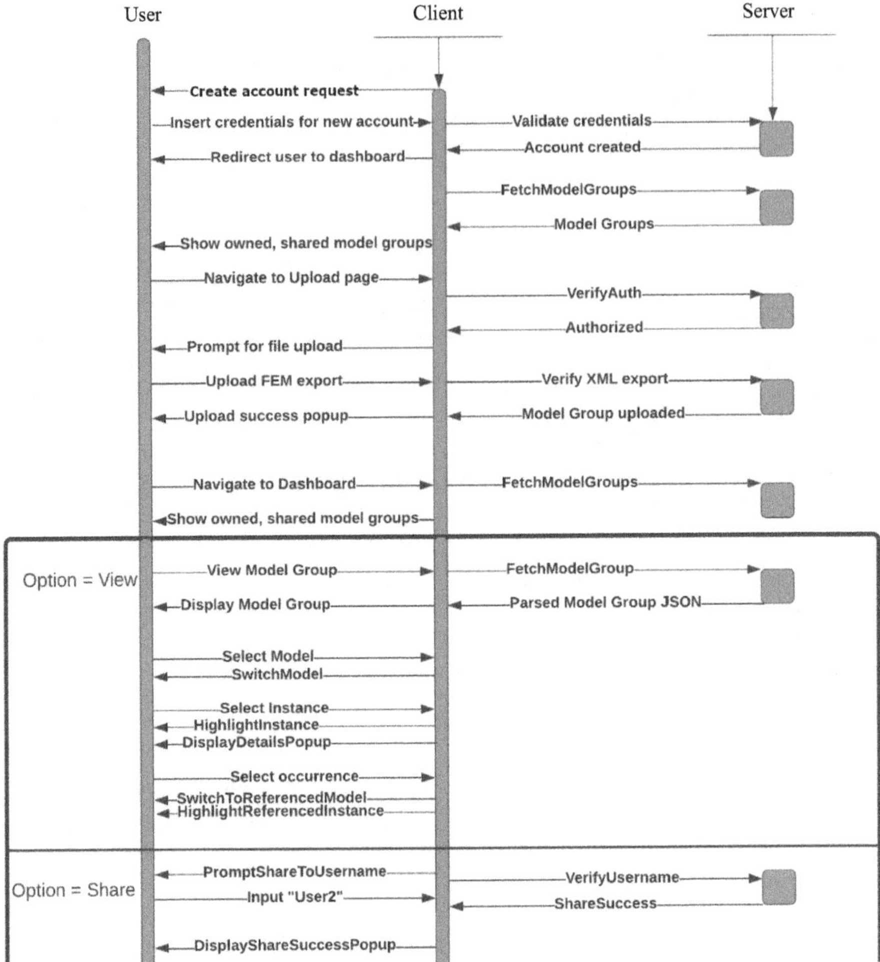

Fig. 7. Interaction between user, client and server

5 Conclusion and Plans for the Future

The FEM toolkit described in the previous section fulfills the goal of giving easy access to the package of FEMs to nontechnical users. The drawback of the current version is that the user cannot leave feedback of any kind. Our plans for the future are to enrich the tool so that a user can leave feedback on any model element using some kind of annotation. This feedback will be possible to import back to the FEM toolkit so that a modeler can see it and make changes accordingly.

Acknowledgments. The FEM view was implemented by the second author as part of his bachelor thesis at the University of Tartu. The first author's work was partly supported by the Estonian Research Council (grant PRG1226).

References

1. Bider, I., Perjons, E., Elias, M., Johannesson, P.: A fractal enterprise model and its application for business development. SoSyM **16**(3), 663–689 (2017)
2. Bider, I., Perjons, E., Klyukina, V.: Tool Support for Fractal Enterprise Modeling. In: Karagiannis, D., Lee, M., Hinkelmann, K., Utz, W. (eds.) Domain-Specific Conceptual Modeling, pp. 205–229. Springer, Cham (2022). https://doi.org/10.1007/978-3-030-93547-4_10
3. ADOxx.org: ADOxx. https://www.adoxx.org. Accessed January 2024
4. Fractalmodel.org: Fractal Enterprise Model. https://www.fractalmodel.org/. Accessed February 2023
5. Langel, S.: FEM. https://github.com/siimlangel/FEM. Accessed January 2024
6. Meta Open Source: React.js. https://react.dev/. Accessed January 2024
7. The OpenJS Foundation: Node.js. https://nodejs.org/en. Accessed January 2024
8. The OpenJS foundation: Express. https://expressjs.com/. Accessed January 2024
9. Prisma Data, Inc.: Prisma. https://www.prisma.io/. Accessed January 2024
10. Passportjs.org: Passport. https://www.passportjs.org/. Accessed January 2024

Ontology-Based Interaction Design for Social-Ecological Systems Research

Max Willis[1(✉)] and Greta Adamo[2]

[1] Bilbao, Spain
maxwillis@humanfactorsinsemantics.net
[2] BC3 - Basque Center for Climate Change, Leioa, Spain
greta.adamo@bc3research.org

Abstract. Contemporary social-ecological systems (SESs) research that supports policy and decision-making to tackle sustainability issues requires interdisciplinary and often multistakeholder synergy. Various frameworks have been developed to describe and understand SESs, each producing different kinds of data and knowledge. The resultant lack of interoperability spurred our development of an ontologically grounded SESs integrated conceptual model. This paper explores the deployment of that model and describes techniques for ontology-based interaction design to clarify notions, align perspectives, negotiate terminologies and semantics in inter- and transdisciplinary collaboration settings. We offer examples of interaction *scripts* that utilise ontologies, discursive artefacts, game and play methods, and report on an exploratory workshop playtest which provided preliminary evidence of the potentials for ontology-based participatory sense-making for knowledge co-production.

Keywords: Ontologies · Social-ecological systems · Interaction design

1 Introduction

Understanding complex social-ecological systems (SESs) to support policy and decision-making requires a synergy between domain-specific knowledge and action in inter- and transdisciplinary settings. Collaboration, participation, and multi-perspective knowledge co-production are central to this endeavour [3,13] for which effective, unambiguous communication between scientists and stakeholders is required [13]. A variety of descriptive and explanatory frameworks have been proposed [3,4] which attempt to capture meanings and structure global and local understandings of SESs, and various engagement techniques have been developed to promote inclusive and equitable SESs analysis and transformation, for example dialogues, collective narratives, and participatory modelling [3,18]. Although these frameworks and activities share the goal of representing and explaining SESs towards sustainable management of human-nature interactions,

M. Willis—Independent researcher.

J. Araújo et al. (Eds.): RCIS 2024, LNBIP 514, pp. 74–82, 2024.
https://doi.org/10.1007/978-3-031-59468-7_9

they produce knowledge following different worldviews, are often difficult to integrate or compare, and can be semantically vague, which hinders interoperability and cross-examination of research data and results [1,3].

In this paper and an accompanying workshop at the Research Challenges in Information Sciences 2024 Conference (see https://humanfactorsinsemantics.net/RCIS2024.html), we explore the potential of ontologies and ontology-based conceptual models to support participatory sense-making and knowledge co-production in sustainability research. The work revolves around an integrated conceptual model and emergent *Social-Ecological SystemS Integrated ONtology* (SESsION) that bridge two well-known SESs frameworks, the social-ecological system framework (SESF) [11] and the ecosystem services (ESs) cascade [14] to clarify key SESs components and integrate them into a unified model using foundational ontologies and related ontological literature [1,2]. The ontology and integrated model, with their unambiguous semantics, ontological clarity and grounding in formal ontologies, align disparate SESs perspectives for understandability and comparability of research, and to facilitate data collection and interpretation.

The SESs integrated conceptual model and underlying ontological basis formalised in SESsION are intended for (i) domain experts to collectively map and express knowledge and support interoperability and comparability with other sustainability efforts, (ii) policy makers and broader communities of practice to access complex SESs knowledge and make sense of local social-ecological scenarios and sustainable management needs, and (iii) developers to design data structures and software alignments that facilitate FAIR datasets for modelling and comparative research. In order to achieve these uses, we design interactions to make sense *of the conceptual model* and render the complexity of SESs entities and relations understandable for researchers and stakeholders, to examine both general and place-based SESs scenarios *through the conceptual model*, and to *interrogate the model itself*, questioning its ontological assumptions and reflexively negotiating its meanings together with communities of knowledge engineers. We approach these tasks through the construction of tangible cognitive artefacts [15,19] and development of participatory sense-making *scripts* [7,10], using game and play interaction design techniques [16,20].

In the following, Sect. 2 briefly reviews the SESs conceptual integration, Sect. 3 covers participatory methods for participatory sense-making, Sect. 4 outlines ontology-based interaction design techniques with three script examples, and Sect. 5 shares some preliminary insights from initial playtesting. Section 6 summarises what has been accomplished and proposes future investigations.

2 Social-Ecological Systems Integrated Perspective

SESs include interdependent *social* and *ecological* components that are often challenging to understand and define [1,6]. Two frameworks that capture SESs knowledge and feature prominently in sustainability research literature are the SESF [11] and the ESs cascade [14]. The former is a product of political science and aims to define a vocabulary for human-nature interactions and common-pool resource management, such as small-scale fisheries. The latter originated from environmental economics, focuses on nature's benefits to humans' well-being, and elaborates ecosystem services and products in terms of benefits and values, such as food and pollination. The two frameworks apply different forms of analysis and produce different kinds of knowledge and data [1–3].

In recognition of the potential benefits of data interoperability and integration of different knowledge sources, such as improving the quality, speed and cohesion of actions, we explore the integration of these two SESs frameworks in [1,2], achieving semantic clarification of central SESF and ESs cascade elements through ontological analysis and conceptual alignment. We reference in particular the Unified Foundational Ontology (UFO) [8] and Descriptive Ontology for Linguistic and Cognitive Engineering (DOLCE) [5]. The resultant SESs integrated model includes four elements from SESF, i.e. *resource*, *resource system*, *actor*, *governance*, and five from the ESs cascade, i.e. ecosystem *structure* and *function*, ecosystem *service*, *benefit* and *value*. For a deep dive into the integrated perspective we redirect the readers to the original articles [1,2] and to SESsION, the latest evolution of which is available for consultation, download, and reuse on GitHub.

3 Participation and the Collective Negotiation of Meanings

Interdisciplinary research on complex, adaptive or evolving social-ecological [3] and socio-technical [9] systems frequently employs participatory practices that are rooted in action research (AR), which proposes qualitative, interpretive inquiry and collaborative methods to engage scientists and stakeholders in cycles of planning, action and reflection [3,9]. AR acknowledges multiple forms of exploratory and explanatory knowledge, and has engendered methods such as participatory rural appraisal (PRA) and knowledge co-production [3,12], user-centered and participatory design [9]. Central to these practices is the collective negotiation of terminologies and meanings, a "figuring things together" ([17] p. 55) wherein agreement is reached on which aspects of an investigative context are most relevant and how those concepts should be articulated and represented. The implications of such participatory sense-making [7] are that meanings emerge from, and are formed in collaborative activities [7,15], for example group- and participatory modelling [10,18] which allow for the *elaboration of existing knowledge* through creating model(s) of x, the enaction of which can facilitate the *co-creation of new knowledge* concerning x [7,10,18]. Collaboratively developed models and other representations, such as rich pictures and causal loops [18], serve as cognitive artefacts, the generation and use of which can, for example, align participants' emergent perspectives [7,15]. The materialities and discursive characteristics of designed things [19] can subsequently foster pluralism and dialogue [20] as people coordinate their actions around, and through those artefacts.

The relations between artefacts, people and practices are linked with human intersubjectivity, language and collaboration (see [15,20]), and their recursive and co-constitutive configurations can be observed and modified through the familiar forms and social coordination of games and play [16]. Game techniques are widely used in participatory research [3,18] and their unique entanglement of material and social worlds are evidenced for example when participants automatically arrange themselves around a gameboard as players, when they collectively acknowledge the gameboard as a map of the game world, and play with game pieces as virtual representations of ourselves. Acceptance of turn-taking, role-play and rules, parts of the social contract that is entered upon agreeing to play a game [16], signals and facilitates willingness to collaborate and the activation of intersubjective linkages at the core of communication [7,15,20].

4 Ontology-Based Interaction Design: Artefacts and Scripts

In this section, we describe an initial approach for an ontology-based interaction design that involves creating an ontology-based conceptual model using a diagramming software, extracting the concepts and their relations, constructing tangible artefacts to represent each component and the whole, and devising interaction scripts [10] using these artefacts and game and play methods to guide knowledge co-production activities. As design strategy, we leverage universally familiar game and play methods that embody collaborative sentiments [7,15] to establish shared space and formalise the social contract of gameplay [16].

The scripts we develop for ontology-based interaction design aim to support participants in (*a*) understanding the ontological concepts and commitments of the model, (*b*) collaboratively constructing meanings and interpreting research and design scenarios guided by the model, and (*c*) interrogating the ontological groundings, assumptions and interpretations of the information artefact/model itself. Scripts include (i) a *name* that frames the engagement, and basic instructions for (ii) setting the *stage* (iii) participants' *roles*, (iv) necessary *props* and (v) procedures or *rules*. Most scripts require a **Mediator** who introduces the activity and initial engagements, and **Players** who actively engage in the activities. Additional roles such as a **Recorder** who takes notes and **Data Entry** who enters data into a tablet spreadsheet may be required. **Spectators** are any observers who are not (yet) engaged in an active role; input from spectators is encouraged, and they can become players at any time. The main **Props** are laminated concept cards which represent each model element and narrow, rectangular tabs for element relations. Other props include game boards (e.g. a large print of the model, henceforth called the big model map, and scenario illustrations) as well as graphical elements, sticky-notes, whiteboard, writing materials, a tablet computer, digital camera and spreadsheet app.

In the following we introduce three scripts for participatory sense-making and knowledge co-production in sustainability research using artefacts derived from the SESs integrated model, including concept cards (e.g. *goal*, *governance*, *natural resource*) and relations tabs (e.g. *participates*, *performed_by*). The design choice of playing cards was to allow participants to dynamically re-arrange model elements, and for game-boards to orientate participants around shared space and activities. Turn-taking and flexible rules were added to structure interactions and scaffold learning, while allowing for interpretation and experimentation. Figure 1 depicts a sample of the props (left) and their use during playtesting (right).

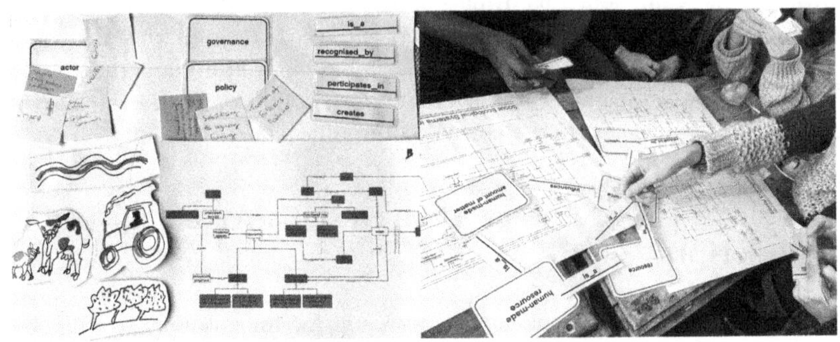

Fig. 1. Sample of SESs props (left), **Relation-Slap!** script test (right).

Relation-Slap! is a script for participants to gain an initial understanding of the model, its elements and their relationships to others. The interaction is inspired by the Japanese card game Menko, in which players slam cards on top of each other in a defined playspace. Props for this activity include the big model map, four model zooms, concept cards and relations tabs. Players sit on the floor or stand at a table around the big model map, with the model zooms at hand. Roles in this script are Mediator and Players, and the activity proceeds as follows:

- Distribute relations tags among players;
- Shuffle and place the big concept cards in the center;
- The Mediator acts as dealer and turns over the top card one by one;
- Each player must "slap!" their relationship tabs between relevant concepts;
- Move the concept cards around to create model snippets;
- Turn-taking may or may not be necessary;
- First player to properly place all of their relationship tabs wins.

Scenario Zoom is a script for domain experts and/or stakeholders to investigate SESs scenarios. Artefacts needed for this activity include scenario illustrations, model zooms, tablet, and spreadsheet, big model map, concept cards, stickynotes and writing materials. Participants sit or stand around a table with the scenario illustration as a game board; the big model map is on hand and model zooms arranged like place mats. Roles include Mediator, Players, Recorder and Data Entry, and the activity unfolds as follows:

- Mediator draws from the stack of large concept cards;
- Players articulate the concept card and its relations referring to model zooms;
- Players write on stickynotes and attach to relevant concept cards;
- Data Entrist adds information to a spreadsheet on the tablet;
- Iterate for 15 min, after which a collective narrative for sustainable policy is produced and written down by the Recorder.

Negotiating commitments is a meta-script to analyse the ontology and conceptual model, which uses the big model map, concept cards, whiteboard, markers and digital camera. The whiteboard is placed flat and participants sit or stand around it, with big model map on hand for reference. No Mediator is required; Players and Recorder do as follows:

- Concept cards are distributed among players;
- Taking turns, place and discus entity cards on the whiteboard;
- Draw lines, arrows and named relations;
- Players take turns to query the ontological commitments, analysis, and alignment with existing ontologies, e.g. "do the relations all make sense?", "How to reconcile the DOLCE ontological commitments?";
- Players reflect upon the understandability and (re)usability of the model, drawing proposed changes on the whiteboard;
- Recorder takes notes and photos of proposed changes and justifications.

5 Preliminary Results

A playtest of **Relation-slap!** and **Scenario Zoom** was enacted in a local cafe with a group (n=5) of post-graduate level sustainability researchers. **Negotiating commitments** was not tested as it requires specific ontology engineering expertise. The authors briefly introduced the SESs integrated model and SESsION ontology, then shared the Mediator role for the exercises, each roughly an hour and a half in duration, after which participants were asked for feedback in the round. Handwritten notes were taken and transcribed post-engagement, with additional insights drawn from researchers' discussions and observations of participant body language and interactions. These texts were examined using discourse analysis techniques, yet due to the small number of participants in a singular engagement, the following insights must be regarded as preliminary.

Signs of alignment between participants can be found by attending to dialogic syntax and resonances in particular, the repetition of utterances and re-elaboration of ideas during the interactions, which signify emergent intersubjectivity [7,20]. One example of resonance is the use of the term "lost" by several participants when describing initial engagement with model complexity, another is their elaboration of the concept of "vision". One participant referred to the model print as a map, stating that with it, "We can see where we are going..." A second suggested, "It's good to have the [scenario] image, so people can see..." and a third shared that Relation-slap! was "...interesting, to get a glimpse of how ontologies work..." These examples offer evidence that participants are engaging with each other's perspectives, and more significantly, articulating how ontology and model can potentially serve as cognitive artefacts, to navigate complexity as an unfamiliar space, and to gain insights from different perspectives.

The discursive dispositions of the artefacts and game interactions were noticeable as the group engaged in animated discussions, commented and disputed each other's actions while animatedly rearranging the cards and relations tabs. During Relation-slap! participants reflexively noted the connection between placing the relations and speaking out their intended logic. One asked if that articulation was a "rule"; another stated, "I think everyone who put the card must explain the card. I placed my last card and there was no discussion, so I started to have doubt in my mind if it was correct." This hints of acceptance that individual contributions could only validated by the group. Participants' body language also provided testimony of enacted social cognition [7]; for example, several times while arranging the artefacts players touched the same card at once and then debated possibilities until agreeing upon the proper configuration. One participant spoke of a plurality that was made possible through the exercises "...maybe we have different opinions. Why? I explain this connection (gesturing to the cards), but another player can explain in a different way."

The social coordination on display can be attributed to well-known aspects of engagement in games as socially structured play, and the positive feedback may simply reflect the novelty effect, as participants had not previously encountered such ontology-based interactions. However it was clear that the ontology was playing a role in the group's emergent worldview. For example, in the scenario exercise, without prompting, participants decided for *organisation* as a *social actor* as per the model, and debated whether *governance* is a *social actor*, or a *role* played by an *organisation*. This spontaneous alignment with the ontological commitments, through an embodied engagement between participants, artefacts and dialogue offers encouraging signs that ontology and interaction design were functioning as coordinating enablers [15].

6 Conclusions and Future Works

This paper presents exploratory research traversing the fields of ontology, conceptual modelling, interaction design and sustainability. It describes a process of transforming an integrated SESs conceptual model and its underlying ontological ground into tangible cognitive artefacts and game and play interactions. Several scripts for engaging groups of scientists and stakeholders in participatory sense-making for knowledge co-production are presented. Results from initial playtests suggest that ontology-based interaction design can indeed guide integrated knowledge co-production for sustainability sciences, and hint at how game and play methods could prove useful for participatory sense-making in the fields of conceptual modelling and applied ontology as well. Future steps include refining the interaction techniques, testing the meta-script, extending and documenting the SESsION ontology.

Acknowledgements. This research is supported by the Basque Government IKUR program Supercomputing and Artificial Intelligence (HPC/AI), the María de Maeztu Excellence Unit 2023-2027 (CEX2021-001201-M) funded by MCIN/AEI /10.13039/501100011033. We thank the playtest participants and the RCIS community for their much-appreciated reviews that helped to improve this work.

References

1. Adamo, G., Willis, M.: Conceptual integration for social-ecological systems. In: Guizzardi, R., Ralyté, J., Franch, X. (eds.) Research Challenges in Information Science. RCIS 2022. LNBIP, vol. 446, pp. 321–337. Springer, Cham (2022). https://doi.org/10.1007/978-3-031-05760-1_19
2. Adamo, G., Willis, M.: The omnipresent role of technology in social-ecological systems: ontological discussion and updated integrated framework. In: Nurcan, S., Opdahl, A.L., Mouratidis, H., Tsohou, A. (eds.) Research Challenges in Information Science: Information Science and the Connected World. RCIS 2023. LNBIP, vol. 476, pp. 87–102. Springer, Cham (2023). https://doi.org/10.1007/978-3-031-33080-3_6
3. Biggs, R., De Vos, A., Preiser, R., Clements, H., Maciejewski, K., Schlüter, M.: The Routledge Handbook of Research Methods for Social-ecological Systems. Taylor & Francis (2021)
4. Binder, C.R., Hinkel, J., Bots, P.W., Pahl-Wostl, C.: Comparison of frameworks for analyzing social-ecological systems. Ecol. Soc. **18**(4), 1–20 (2013)
5. Borgo, S., et al.: DOLCE: a descriptive ontology for linguistic and cognitive engineering. CoRR abs/2308.01597 (2023)
6. Colding, J., Barthel, S.: Exploring the social-ecological systems discourse 20 years later. Ecol. Soc. **24**(1), 1–10 (2019)
7. De Jaegher, H., Di Paolo, E.: Participatory sense-making: an enactive approach to social cognition. Phenomenol. Cogn. Sci. **6**, 485–507 (2007)
8. Guizzardi, G., Benevides, A.B., Fonseca, C.M., Porello, D., Almeida, J.P.A., Sales, T.P.: UFO: unified foundational ontology. Appl. Ontol. **17**(1), 167–210 (2022)
9. Hayes, G.R.: The relationship of action research to human-computer interaction. ACM Trans. Comput. Human Interact. (TOCHI) **18**(3), 1–20 (2011)
10. Hovmand, P.S., Andersen, D.F., Rouwette, E., Richardson, G.P., Rux, K., Calhoun, A.: Group model-building 'scripts' as a collaborative planning tool. Syst. Res. Behav. Sci. **29**(2), 179–193 (2012)
11. McGinnis, M.D., Ostrom, E.: Social-ecological system framework: initial changes and continuing challenges. Ecol. Soc. **19**(2), 1–28 (2014)
12. Miller, C.A., Wyborn, C.: Co-production in global sustainability: histories and theories. Environ. Sci. Policy **113**, 88–95 (2020)
13. Newell, B.: Simple models, powerful ideas: towards effective integrative practice. Glob. Environ. Chang. **22**(3), 776–783 (2012)
14. Potschin, M., Haines-Young, R., et al.: Defining and measuring ecosystem services. In: Routledge Handbook of Ecosystem Services, pp. 25–44 (2016)
15. Stahl, G.: Meaning and interpretation in collaboration. In: Wasson, B., Ludvigsen, S., Hoppe, U. (eds.) Designing for Change in Networked Learning Environments. Computer-Supported Collaborative Learning, vol. 2, pp. 523–532. Springer, Dordrecht (2003). https://doi.org/10.1007/978-94-017-0195-2_62
16. Stenros, J.: In defence of a magic circle: the social, mental and cultural boundaries of play. Trans. Digital Games Res. Assoc. **1**(2), 1–39 (2014)

17. Suchman, L.: Configuration. In: Inventive Methods, pp. 48–60. Routledge (2012)
18. Voinov, A., et al.: Tools and methods in participatory modeling: Selecting the right tool for the job. Environ. Model. Softw. **109**, 232–255 (2018)
19. Wakkary, R., Odom, W., Hauser, S., Hertz, G., Lin, H.: Material speculation: actual artifacts for critical inquiry. In: 5th Decennial Aarhus Conference on Critical Alternatives August (AA 2015), pp. 97–108. Aarhus University (2016)
20. Willis, M.: On agonism and design: dialogues between theory and practice. Ph.D. thesis, University of Trento (2019)

Scriptless and Seamless: Leveraging Probabilistic Models for Enhanced GUI Testing in Native Android Applications

Olivia Rodríguez-Valdés[3(✉)], Kevin van der Vlist[1], Robbert van Dalen[1], Beatriz Marín[2], and Tanja E. J. Vos[2,3]

[1] ING Bank, Amsterdam, The Netherlands
[2] Universitat Politècnica de València, Valencia, Spain
[3] Open Universiteit, Heerlen, The Netherlands
orv@ou.nl

Abstract. The growing mobile app market demands effective testing methods. Scriptless testing at the Graphical User Interface (GUI) level allows test automation without traditional scripting. Nevertheless, existent scriptless tools lack efficient prioritization and customization of oracles and require manual effort to add application-specific context, hindering rapid application releases. This paper presents Mint as an alternative tool that addresses these drawbacks. Preliminary results indicate its capability to detect accessibility problems.

Keywords: Mobile apps · test automation · scriptless testing · accessibility oracle

1 Introduction

With over 6.3 billion smartphone users worldwide [1], the demand for effective Android and iOS app testing methods is rising [2] due to continuous app usage and smartphone adoption growth [3]. GUI-level testing, mirroring the user experience, has seen a shift from thorough but costly manual testing [4] to automated methods that, despite speeding up the process, often struggle with constant updates and the challenge of covering all GUI paths in the AUT [5].

Scriptless testing [6] offers automation without the reliance on traditional scripting. It utilizes agents to choose optimal test actions and has been demonstrated to be complementary [2,7] to scripted methods by reducing maintenance and covering more paths than scripted approaches [6]. Existing scriptless tools for Android lack efficient prioritization and advanced oracles to detect problems in the AUT and need significant manual input for specific AUT information.

This paper presents Mint: an implementation of the TESTAR scriptless testing approach [6] for Android that solves some of these drawbacks. The implementation encompasses: (1) customizable rules, (2) probabilistic exploration to improve coverage, (3) composable oracles, (4) effortless integration, and (5) improved reporting. Furthermore, the paper describes a preliminary empirical

J. Araújo et al. (Eds.): RCIS 2024, LNBIP 514, pp. 83–91, 2024.
https://doi.org/10.1007/978-3-031-59468-7_10

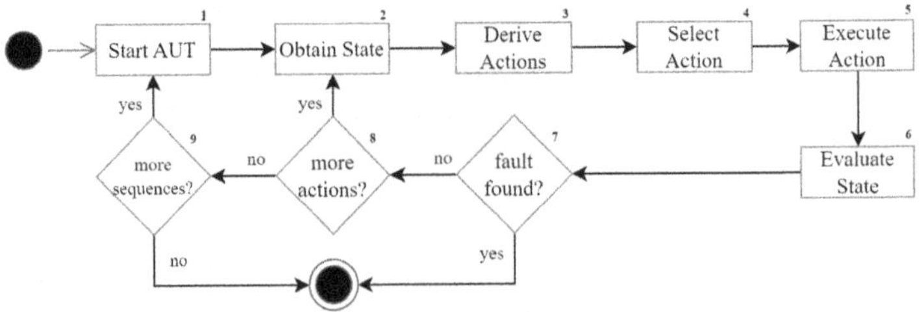

Fig. 1. Scriptless testing logical execution flow.

comparison with two relevant scriptless testing tools applied to one mobile application. The selected testing tools are Testar [6] and DroidBot [8].

The contributions of this paper are: (1) a novel scriptless testing tool for Android that uses a probabilistic model and composable oracles for the selection of actions, and (2) a preliminary comparison with Testar and DroidBot in terms of effectiveness and efficiency. The paper is structured as follows. Section 2 presents the main characteristics of scriptless GUI testing, and a brief discussion of tools for Android testing. Section 3 introduces Mint along with its key features. Section 4 presents preliminary results from the initial comparative experiment. Section 5 presents our main conclusions and future research directions.

2 Scriptless GUI Testing

Scriptless testing involves generating and executing tests on-the-fly, identifying the current state of the AUT, and selecting actions using Action Selection Mechanisms (ASM) to transition to the next state. This approach automatically identifies possible actions based on the GUI structure. ASMs include random choices or more intelligent strategies.

Outlined in Fig. 1, the process begins by starting the AUT and preparing it for interaction. Next, the AUT is thoroughly inspected to understand its current state and components (Step 2). Next, actions available in this state are identified (Step 3). From this derived list of actions, an appropriate one is chosen using an ASM (Step 4) and then executed on the AUT (Step 5). After each action or test sequence completion, test oracles are employed to check for failures (Step 6). If no faults are detected, steps 2 to 6 are repeated (Step 7) until the desired test sequence length is achieved (Step 8). Finally, the tool either concludes and exits gracefully upon reaching the predetermined number of test sequences or restarts the process from Step 1 (Step 9).

2.1 Existing Tools for Mobile Applications

Our review of scriptless testing tools for Android apps showed diverse approaches. Tools like Testar [6], Dynodroid [9] and DroidBot [8] use **random action selection**, with DroidBot adding a **model-based approach** for state recognition. Stoat [10] and APE [11] also use model-based strategies for unexplored areas.

Advanced **ASM** are employed in other tools. Sapienz [12] uses fuzzing and search-based methods with evolutionary algorithms. Humanoid [13] applies deep-learning to mimic human interactions, while ARES [14] uses deep-learning for better exploration strategies. QTesting [15] and AimDroid [16] use reinforcement learning, the former for prioritizing unfamiliar states and the latter for predicting events likely to trigger new activities or crashes. ComboDroid [16] uses combinatorial exploration for identifying unvisited states, and RegDroid [17] focuses on finding functional bugs via differential regression testing.

While diverse, these tools face several limitations: inefficient prioritization in random testing, extensive training for reinforcement learning tools, and most depend on implicit oracles for detecting issues. Many are not maintained, leading to compatibility issues with newer Android versions. Finally, these tools mainly focus on code coverage as an effectiveness measure, lacking detailed fault reports.

3 MINT

Mint[1] is a scriptless testing tool that uses a probabilistic exploration approach augmented by customizable rules to identify and (de)prioritize GUI interactions during action selection. Once configured, Mint autonomously explores native Android applications, obviating the need for manual script development and maintenance. Unlike scripts, which require detailed programming to define every test case, Mint's rules allow testers to specify testing criteria and priorities more abstractly and intuitively, facilitating rapid adaptation to application changes.

Mint employs oracles to verify the AUT's intended behavior, producing verdicts that record specific aspects of the AUT, which facilitates a comprehensive evaluation of the AUT. An interactive reporting tool accompanies Mint, providing detailed analysis of system interactions and insights into the AUT's behavior, crucial for enhancing the quality and Customer Experience (CX).

Mint uses the Espresso API for UI identification and interaction in Android, automated GUI tests. Mint starts by accessing the top-level View container, encompassing all GUI elements, and recursively traverses the GUI hierarchy, aggregating View elements to form a concrete state representation saved as an XML document. An example of this XML state representation, featuring one container layout and three widgets, is shown in Fig. 2. Mint integrates as a plugin in testing pipelines, similar to unit or integration tests, enabling seamless inclusion of exploratory GUI testing in Android application development workflows. MINT introduces three key features: customizable rules, probabilistic exploration, and composable oracles.

3.1 Customizable Rules

Mint operates based on predefined rules dictating interactions with the AUT, like clicks and text inputs. Each rule has a relative importance, forming a model that

[1] https://github.com/ing-bank/mint.

```
<View class="Layout" ...>

 <View class="Checkbox"
       id="like" .../>

 <View class="TextField"
       id="comment" .../>

 <View class="Button"
       id="done" .../>


</View>
```

Fig. 2. State representation

```
<View class="Layout" ...>
 <View class="Checkbox" id="like" ...>
  <action type="click" prio="1.0" .../>
 </View>
 <View class="TextField" id="comment" ...>
  <action type="text" prio="2.0" value="text" .../>
  <action type="text" prio="1.0" value="1282" .../>
 </View>
 <View class="Button" id="done" ...>
  <action type="click" prio="2.0" .../>
  <action type="multiplicative" prio="3.0" .../>
 </View>
</View>
```

Fig. 3. Annotated state representation

Table 1. Predefined rules provided by MINT

Generic Rules		
Functionality	Description	Rule Names
Click and Scrolling Rules	Rules for clicking widgets based on visibility, clickability, position, and presence in pop-ups or adapter views. Includes prioritising and deprioritising clicks.	simpleClickableRule, scrollingClickableRule, clickableRuleForItemWithTag, clickableRuleBasedOnPositionInViewHierarchy, clickableRuleBasedOnPositionInViewHierarchy-ForPopupItem, deprioritizeClickingOnPopupItemOnCurrent-Root, clickableRuleForAdapterViewItems, clickableRuleForSpinnerItems, spinnerSimpleClickDeprioritizeRule, adapterViewSimpleClickDeprioritizeRule
Device Rules	Rules for changing device orientation and theme to test UI responsiveness.	deviceRotationRule, deviceThemeRule
Selection and Input Rules	Rules for navigating and inputting in pagers, TimePickers, and DatePickers.	scrollingPagerRightRule, scrollingPagerLeftRule, timePickerInputRule, datePickerInputRule, clickableRuleBasedOnPositionInViewHierarchy-ForBottomSheet, defaultPreviousActionDeprioritizeRule
Specific Rules		
Input Rules	Generate text inputs for various types: UTF8, generic text, multiline, email, numbers (standard, decimal, signed), person name, URI, phone number, postal address, date, and time.	defaultUTF8InputRule, defaultTextInputRule, defaultMultilineTextInputRule, defaultEmailAddressInputRule, defaultNumberInputRule, defaultDecimalNumberInputRule, defaultSignedNumberInputRule, defaultPersonNameInputRule, defaultUriRule, defaultPhoneNumberInputRule, defaultPostalAddressInputRule, defaultDateInputRule, defaultTimeInputRule
Click Deprioritizing Rules	Rules to de-prioritize clicking on uneditable text fields, text elements, and based on position	defaultUneditableTextClickDeprioritizeRule, defaultTextClickDeprioritizeRule, defaultTextClickAtPositionDeprioritizeRule

intelligently guides Mint through various AUT states without prior knowledge. Rules are defined as tuples $R = (P, A, \pi)$. The predicate function $P : S \rightarrow \{0, 1\}$ maps states $s \in S$ to a binary set, with $P(s) = 1$ indicating rule applicability in state s. The set A encompasses all executable actions. Priority $\pi \in \mathbb{R}$ assigns each rule's relative importance, influencing the execution order in Mint's decision-making. Rules include attributes like *name*, *description*, and *modifier*.

Table 2. Implemented oracles provided by MINT

Category	Oracle Names
Stability, Performance	AndroidDeviceOracle, AndroidLogOracle, CrashOracle
Accessibility	ClassNameCheckOracle, ClickableSpanCheckOracle, DuplicateClickableBoundsCheckOracle, DuplicateSpeakableTextCheckOracle, EditableContentDescCheckOracle, ImageContrastCheckOracle, LinkPurposeUnclearCheckOracle, RedundantDescriptionCheckOracle, SpeakableTextPresentCheckOracle, TextContrastCheckOracle, TextSizeCheckOracle, TouchTargetSizeCheckOracle, TraversalOrderCheckOracle, UnexposedTextCheckOracle

The *modifier* adjusts a rule's priority, allowing fine-tuning for its decision-making impact. For instance, a rule with an initial priority of $\pi = 0.5$, when modified by a factor of 2, changes its priority to $\pi = 1.0$, thus doubling its importance.

Mint categorizes its rules into three types:

- Generic rules: applicable to all AUT, e.g., interacting with clickable elements.
- Specific rules: tailored for particular application types, addressing unique scenarios such as inputting email addresses in relevant email input fields.
- Domain-Specific rules: designed for internal use with specific account numbers or identifiers.

Table 1 shows all existing rules provided by Mint. These include navigation rules for actions like scrolling to and clicking unseen widgets, or deprioritizing previously executed actions. Each deprioritizing rule has a multiplicative modifier to lower its priority. Mint also features input rules to generate diverse inputs (e.g., emails, names, dates, postal codes) for different test scenarios. For example, the generic rule *simpleClickableRule* defined as:

```
GenericRule(pred = xpred(".[@isClickable='true' and @isDisplayed='true']"),
            action = Action.CLICK,   prio = 0.5)
```

uses Espresso to check if an element is clickable and visible, then applies a click action with a set priority. Specific input types are defined using regular expressions for data like emails or numbers. To generate realistic, non-sensitive data such as phone numbers or postal codes, Mint uses the JavaFaker library.

Algorithm 1. Action Selection Mechanism of MINT

```
procedure SelectAction(actionRules)
    pairs ← []
    prioritySum ← 0
    for rule in actionRules do
        modPriority ← (rule has modifier)? ModifyPriority(rule) : rule.Priority
        pairs ← pairs + (rule.action, modPriority)
        prioritySum ← prioritySum + modPriority
    dice ← RandomFloat() × prioritySum
    for (action, modPriority) in pairs do
        if (accumPriority ← accumPriority + modPriority) ≥ dice then
            return action
```

3.2 Probabilistic Exploration

Mint links actions to nodes in its state representation. The process involves iterating over each rule, traversing widget nodes within the state, and attaching the rule's action to nodes satisfying its predicate. This is shown in Fig. 3, where actions and their attributes are annotated.

The ASM, explained in Algorithm 1 , works in two phases. First, it calculates priorities for all applicable rules, considering modifiers that can alter these priorities, resulting in a set of (*rule, modified priority*) pairs. The next action is selected by generating a random number within the range of cumulative priorities and choosing the action with a higher number.

3.3 Composable Oracles

Mint features an oracle module to evaluate different aspects of the AUT, categorizing oracles based on specific testing goals such as Accessibility, Internationalisation, Performance, Stability, Aesthetics, and Miscellaneous. Each oracle's structure, defined through an interface, includes category, probe, and evaluation function. Probes act as data sources, enabling oracles to assess the AUT. For example, the AndroidLogOracle uses system logs as a probe to detect faults.

Accessibility-focused oracles form a significant part of Mint, ensuring compliance with accessibility standards [18] by evaluating text readability, image contrast, and other factors for optimal CX. Existing oracles, listed in Table 2, also include *CrashOracle* for crash detection and *AndroidDeviceOracle* for monitoring CPU and memory usage.

Mint's oracle framework is designed for extensibility, allowing customization and combination of oracles for comprehensive test suites. This flexibility facilitates a versatile testing environment. For example, testers can define a rule using the AndroidLogOracle for fault checking while excluding the AndroidDeviceOracle for memory monitoring, as shown in the following code:

```
MINTRule(DefaultBuilder.withOracle(AndroidLogOracle)
                       .withoutOracle(AndroidDeviceOracle).build())
```

Table 3. Comparison of Testing Tools on Various APKs

AUT	LOC	Metric	Droidbot	Testar	Mint
Amaze File Manager	84247	ACC	15.6%	**23.7%**	22.5
		Failures	0	0	46
Arity	5197	ACC	26.0%	**66.6%**	40.7%
		Failures	0	0	5

3.4 Seamless Integration

Mint offers a plugin for seamless integration into Android testing workflows, complementing standard unit and integration tests. This plugin enables easy addition of Gradle tasks for data collection and report generation within the Android testing framework. Since Gradle is the default automation tool in Android Studio, integrating Mint with Gradle-based projects is straightforward, requiring just an addition to the Gradle configuration. Mint tests can be defined and run akin to standard Android testing, as shown in the example code:

```
@org.junit.Test
fun mintExploratoryTestRun() {
    MINTRule().explore() }
```

3.5 Reporting the Results

Mint records testing data in XML format. The mentioned plugin includes a reporting task that parses this XML, generating an HTML overview of oracle outputs, individual test sequence pages, and associated screenshots. It also features a search function for AUT elements via XPath, highlighting their locations in the screenshots. This reporting structure, organized into HTML pages, significantly improves the clarity and analysis of test results (Table 3).

4 Preliminary Evaluation

A preliminary evaluation to compare Mint with actively maintained random testing tools reported in Sect. 2.1 was done. To do that, we select DroidBot and Testar. The AUTs were selected randomly from F-Droid, a platform for distributing free and open-source Android apps. The first AUT, *Amaze File Manager*, is an advanced file explorer that allows different operations over the Android file system. Arity is a scientific calculator with function graphing.

A test run consists of testing the AUT for 300 test actions. Ten test runs were executed for every tool, and the final average code coverage (ACC) was obtained. Our comparative analysis of testing tools showed a nuanced distinction in their bug detection capabilities. Mint uniquely excelled in identifying specific types of accessibility issues. This suggests that Mint's specialized focus on certain failure categories, like accessibility, adds a valuable dimension to the testing landscape that the existing tools have not addressed (see Sect. 2). Moreover,

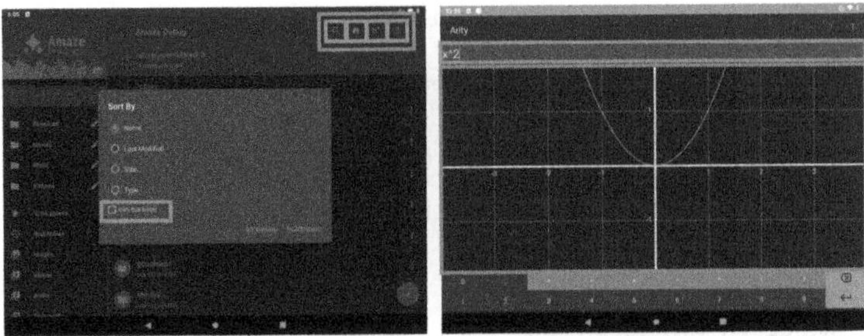

Fig. 4. Sample of accessibility issues detected by Mint Orange: Touch widget size is not large enough. Blue: Widget without speakable text. (Color figure online)

multiple exceptions obtained through the Android Log Oracle were discarded as false positive bugs, except for one identified in Arity.

Mint found 2 different types of accessibility problems on multiple widgets: not speakable text and touch target size not large enough, detected by oracles *SpeakableTextPresentCheckOracle* and *TouchTargetSizeCheckOracle* respectively. Figure 4 depicts a sample of states where such accessibility issues were found. No faults were detected by DroidBot or Testar.

Mint's lower coverage can be explained by its requirement for a manual fine-tuning of the rules, as it is designed for use by testers of the AUT itself, leveraging their knowledge of its specific nature. This contrasts with the plug-and-play approach of the existing tools, which require minimal setup. However, Mint can more effectively balance the trade-off between code coverage and fault detection. This reflects the understanding in software testing that code coverage is a helpful but not definitive indicator of test quality [19].

5 Conclusions and Future Work

This paper presents Mint, a scriptless testing tool for Android applications, which allows the definition of composable oracles and uses probabilistic exploration. Preliminary results show that Mint fills a critical gap in detecting faults like accessibility issues, which are increasingly important in creating inclusive and user-friendly applications. Therefore, rather than viewing Mint's performance in isolation, it should be considered as part of a diverse toolkit, hence its seamless design for easy integration with standard Android testing frameworks.

In future work, we plan to perform a larger empirical evaluation to compare Mint with the tools of the state of the art in terms of the effectiveness measured with the faults detected and coverage, the efficiency, and the perceived satisfaction of the ease of use and installation of the tool. Recognizing the potential challenges of manual fine-tuning in Mint's rule-based approach, we are exploring the integration of machine learning techniques to automate rule configuration.

Acknowledgements. This project was done within the context of the AUTOLINK project, Automated Unobtrusive Techniques for LINKing requirements and testing in agile software development (19521).

References

1. Geiger-Prat, S., Marín, B., España, S., Giachetti, G.: A GUI modeling language for mobile applications. In: 9th RCIS, pp. 76–87. IEEE (2015)
2. Jansen, T., et al.: Scriptless GUI testing on mobile applications. In: IEEE QRS (2022)
3. Buildfire: Mobile app download statistics & usage statistics (2021). https://buildfire.com/app-statistics/. Accessed 26 Nov 2023
4. Asfaw, D.: Benefits of automated testing over manual testing. Int. J. Innov. Res. Inf. Secur. **2**(1), 5–13 (2015)
5. Coppola, R., Raffero, E., Torchiano, M.: Automated mobile UI test fragility: an exploratory assessment study on android. In: INTUITEST. ACM (2016)
6. Vos, T.E., Aho, P., Pastor Ricos, F., Rodriguez-Valdes, O., Mulders, A.: TESTAR-scriptless testing through graphical user interface. Softw. Test. Verif. Reliab. **31**(3), e1771 (2021)
7. Bons, A., Marín, B., Aho, P., Vos, T.E.: Scripted and Scriptless GUI testing for web applications: an industrial case. In: IST 2023 (2023)
8. Li, Y., Yang, Z., Guo, Y., Chen, X.: Droidbot: a lightweight UI-guided test input generator for android. In: IEEE/ACM ICSE-C. IEEE (2017)
9. Machiry, A., Tahiliani, R., Naik, M.: Dynodroid: an input generation system for android apps. In: Proceedings de ESEC/FSE 2013 (2013)
10. Su, T.: FSMdroid: guided GUI testing of android apps. In: IEEE/ACM ICSE-C (2016)
11. Gu, T., et al.: Practical GUI testing of android applications via model abstraction and refinement. In: IEEE/ACM 41st ICSE, pp. 269–280 (2019)
12. Mao, K., Harman, M., Jia, Y.: Sapienz: multi-objective automated testing for android applications. In: ISSTA 2016 (2016)
13. Li, Y., Yang, Z., Guo, Y., Chen, X.: Humanoid: a deep learning-based approach to automated black-box android app testing. In: 2019 34th IEEE/ACM International Conference on Automated Software Engineering (ASE) (2019)
14. Romdhana, A., Merlo, A., Ceccato, M., Tonella, P.: Deep reinforcement learning for black-box testing of android apps. In: ACM TOSEM 2022 (2022)
15. Pan, M., Huang, A., Wang, G., Zhang, T., Li, X.: Reinforcement learning based curiosity-driven testing of android applications. In: 29th ACM SIGSOFT International Symposium on Software Testing and Analysis (2020)
16. Gu, T., et al.: Aimdroid: activity-insulated multi-level automated testing for android applications. In: International Conference on Software Maintenance and Evolution (2017)
17. Xiong, Y.et al.: An empirical study of functional bugs in android apps. In: ACM SIGSOFT 2023 (2023)
18. Android: Android accessibility overview. Accessed 26 Nov 2023. https://developer.android.com/guide/topics/ui/accessibility
19. Myers, G.J., Badgett, T., Thomas, T.M., Sandler, C.: The Art of Software Testing, vol. 2. Wiley Online Library, Hoboken (2004)

Identifying Relevant Data in RDF Sources

Zoé Chevallier[1,2(✉)], Zoubida Kedad[1], Béatrice Finance[1], and Frédéric Chaillan[2]

[1] DAVID Lab, University of Versailles Paris-Saclay, Versailles, France
{zoe.chevallier,zoubida.kedad,beatrice.finance}@uvsq.fr
[2] Grand Paris Sud, Evry-Courcouronnes, France
{z.chevallier,f.chaillan}@grandparissud.fr

Abstract. The increasing number of RDF data sources published on the web represents an unprecedented amount of information. However, querying these sources to extract the relevant information for a specific need represented by a target schema is a complex task as the alignment between the target and the source schemas might not be provided or incomplete. This paper presents an approach which aims at automatically populating the classes of a target schema. Our approach relies on a semi-supervised learning algorithm that iteratively identifies instance patterns in the data source that represent candidate instances for the target schema. We present some preliminary experiments showing the effectiveness of our approach.

Keywords: RDF data sources · Target Schema Instantiation · Web Data extraction · Semi-supervised learning

1 Introduction

The web represents a huge space of available data from which various applications can extract meaningful knowledge. However, finding relevant data for a specific need is not a trivial task. It requires the exploration of these data sources, which can be time-consuming.

The specific needs of an application can be described by a target schema. Our goal is to identify the candidate instance patterns for the classes of this target schema in a given data source. In our work, we focus on datasets expressed in RDF (Resource Description Framework)[1], the standard language proposed by the W3C to describe resources on the web.

This paper presents an approach that determines if a source contains data that are relevant to a given target schema by extracting a set of candidate instance patterns. Each pattern is a property set describing some candidate instances in the data source. Our approach relies on a semi-supervised algorithm that iteratively compares the source entities to both the target classes and the candidate instances already identified.

[1] https://www.w3.org/RDF/.

J. Araújo et al. (Eds.): RCIS 2024, LNBIP 514, pp. 92–99, 2024.
https://doi.org/10.1007/978-3-031-59468-7_11

Our main contribution is the identification of candidate instance patterns for a given target schema in an RDF dataset when the alignment between the source and the target schemas is missing or incomplete. These patterns can be used to generate the SPARQL queries to populate the target schema.

The paper is organized as follows. Section 2 formalizes the problem statement. Section 3 presents our classification model. Section 4 describes the process of generating class descriptions, Sect. 5 presents some experiments, Sect. 6 presents some related works, and finally, Sect. 7 concludes the paper.

2 Problem Statement

Let us consider a target schema T. We aim to instantiate this target schema by finding the relevant patterns representing candidate instances in RDF data sources.

An RDF data source is a set of triples D ⊆ (R ∪ B) × P × (R ∪ B ∪ L), where R, B, P, and L represent respectively the resources set, the blank-node set, the property set and the literal set. In a triple <s, p, o>, s is a resource that is the subject of the triple, p is a property describing the resource s, and o is the object of the triple and represents the value of the property p for the resource s. An RDF data source contains both the data and the schema describing these data. In an RDF data source, we define an entity as follows.

Definition (Entity): An entity e in an RDF data source S is a resource that is not a class, a property, a literal, or a blank node.

An entity e is characterized by a set of properties, denoted $Prop(e)$, such that:

$$Prop(e) = \{p \mid < e, p, * > \in S\} \tag{1}$$

A target schema is defined using RDFS[2] and OWL[3]. A class C is represented in the target schema T by a triple $< C, rdf{:}type, rdfs{:}Class >$. Each property p of C is represented in T by a triple $< p, rdfs{:}domain, C >$. The property set of a given class C is denoted $Prop(C)$, and it is such that:

$$Prop(C) = \{p \mid < p, rdfs{:}domain, C > \in T\} \tag{2}$$

Candidate instance patterns in the data source can be identified by computing the similarity between source entities and the target class C_T, i.e., $Prop(C_T)$ and $Prop(e)$, using similarity measures such as the Jaccard index [4] or the overlap coefficient for example.

If the alignment between the target and the source schemas are provided, identifying candidate instances from the source entities is straightforward. If a target class C_T is equivalent to a source class C_S, then all the instances of C_S are candidate instances of the target class C_T. However, these correspondences are not always provided. Besides, the schema in RDF sources is only descriptive, and

[2] https://www.w3.org/TR/rdf12-schema/.
[3] https://www.w3.org/TR/owl-ref/.

an instance can be characterized by a property set that is different from the one defined in the schema for its class. Therefore, new candidate instances should be identified based on the target schema, but also on the candidate instance patterns already identified.

Our problem can be stated as follows: given a class C defined in the target schema and an RDF data source S, how to identify in S the existing patterns that represent the candidate instances for C. We consider that a pattern is a set of properties describing at least one entity in the data source. These patterns could be later used to automatically generate the extraction queries.

3 Classification Model

In order to determine if an entity e is a candidate instance for a class C, e can be compared to the definition of the class C in the target schema, which consists in comparing the property sets $Prop(e)$ and $Prop(C)$. We introduce the notion of candidate instance descriptions, representing a set of properties describing candidate instances of C in the data source S. It is defined as follows.

Definition (Candidate Instance Descriptions): Let C be a target class and I_C its set of candidate instances. A candidate instance description $CID_i(C)$ of C is a set of properties such that there is a candidate instance e in I_C and $CID_i(C)$ $= Prop(e)$. The set of all the class descriptions of C is denoted $Desc_{CI}(C)$.

To determine if an entity e is a candidate instance for a target class C, we consider two types of similarities: schema-based similarity, which compares the property set of C as defined in the schema to the property set of e, and instance-based similarity, which compares the property set of e to the candidate instances of C, which are represented by the set of class descriptions of C. The final similarity score $Sim(e, C)$ is computed as the maximum value between schema-based and instance-based similarities, which are defined hereafter.

Definition (Schema-based Similarity): Schema-based similarity between an entity e and a class C is based on the Jaccard index [4]:

$$Sim_S(e, C) = \frac{|Prop(e) \cap Prop(C)|}{|Prop(e) \cup Prop(C)|)} \tag{3}$$

Definition (Instance-based similarity): Instance-based similarity is the maximum value of the similarity between e and each class description $CD_i(C)$ in $Desc_C$. It is computed based on the Jaccard index [4].

$$Sim(e, CD_i(C)) = MAX(\frac{|Prop(e) \cap CD_i(C)|}{|Prop(e) \cup CD_i(C)|}, \text{where}\, CD_i(C) \in Desc_C) \tag{4}$$

If the similarity between an entity e and a class C is higher than a similarity threshold denoted τ, i.e., if $Sim(e, C) \geq \tau$, then $Prop(e)$ is a candidate instance

Algorithm 1. Identifying Candidate Instance Descriptions for a Target Class C

Input: RDF data source (S), target schema (T), target class (C), similarity threshold (τ)
Output: $Desc_{CI}(C)$ set of candidate descriptions for C

1: $Desc_{CI}(C) \leftarrow \emptyset$ ▷ initialization of the candidate instance descriptions (classifier)
2: **for all** C_S equivalent to C **do** ▷ schema-alignment based
3:     **for all** e instance of C_S **do**
4:         $Desc_{CI}(C) \leftarrow Desc_{CI}(C) \cup Prop(e)$
5: **for all** e entity in S **do** ▷ target class definition based
6:     **if** $Sim_S(e, C) \geq \tau$ **then**
7:         $Desc_I(C) \leftarrow Desc_I(C) \cup Prop(e)$
8: condition ← True ▷ iterations until no new candidate instances identified
9: **while** condition **do**
10:     $\Delta I_C \leftarrow \emptyset$
11:     **for all** e entity of S $\wedge\, Prop(e) \notin Desc_C$ **do** ▷ prediction of entity type
12:         **if** $Sim(e, C) \geq \tau$ **then**
13:             Add e to ΔI_C
14:     **for all** entity e in ΔI_C **do** ▷ update of the classification model
15:         $Desc_{CI}(C) \leftarrow Desc_{CI}(C) \cup Prop(e)$
16:     **if** ΔI_C is empty **then** ▷ checking of the iteration condition
17:         condition ← False
18: **Return** $Desc_{CI}(C)$

description of C. This similarity is computed as the maximum value between the schema-based and the instance-based similarity, as follows:

$$Sim(e, C) = MAX(Sim_S(e, C), Sim_I(e, C)) \tag{5}$$

4 Generating Candidate Instance Descriptions

To determine the candidate instance descriptions for a class in the target schema from the entities in a data source, we propose to adapt an iterative semi-supervised learning algorithm, the Self Training algorithm [11].

Algorithm 1 describes the identification of candidate descriptions from the data source of a target class C based on Self-Training [11]. It uses the classification model introduced in Sect. 3.

If some correspondences between the source and the target schema are provided, we compute the set of candidate instance descriptions based on the schema alignment (lines 2–4). Then, the entities are compared to the property set of C, *Prop*(C). If the similarity between e and C is higher than a threshold τ, their property sets are considered as candidate descriptions (lines 5–7). For each entity e, if the similarity between e and the target class C is higher than a threshold τ, then e is added to the set of candidate instances of the current iteration, ΔI_C (lines 11–13).

For each candidate instance identified during the iteration, $e \in \Delta I_C$, $Prop(e)$ is added to the descriptions of C, $Desc_{CI}(C)$ (lines 14–15).

The algorithm iterates while new candidate instance descriptions are identified. The process terminates when no new candidate description is discovered during an iteration, as expressed by the condition in lines 16–17.

Using our algorithm, a given entity can be a candidate instance for several target classes. Indeed, assume that an entity e is a candidate instance for a class C_i; if e is compared to class C_j and has a similarity score $Sim(e, C_j)$ greater than the similarity threshold, then e is also a candidate instance for C_j.

5 Experiments

The datasets used to evaluate our approach have been extracted from DBpedia[4]. Each dataset corresponds to one class, and we randomly collected instances of this class with all their associated triples. We have extracted three datasets and presented their characteristics in Table 1.

Table 1. Description of the Datasets

	Number of entities	Number of distinct properties	Average number of properties per entity	Number of distinct patterns
Book	10000	1635	12	5648
Publisher	3883	534	7	2042
Writer	10000	737	11	7271

The instances extracted are typed and the type declarations are considered as the ground truth. We have then removed 80% of the type declarations, and 20% of the instances of each class remained typed. This is close to what can be observed on real data sources, for example, according to Fernandez et al. [3], DBpedia contains 16% of typed entities.

In the datasets, we have excluded the properties that are part of the RDF, RDFS, or OWL vocabularies. These properties can be used to describe any class, and therefore should not be considered in the similarity computation to avoid distortion in the result.

The target schema has been generated using a large language model, ChatGPT[5]. We have issued a query for a target schema composed of Books, Writers, and Publishers and their characteristics. We have assumed the following correspondences: (i) the target class **Book** is equivalent to the source class **Book**, (ii) The target class **Publisher** is equivalent to the source class **Publisher**, and (iii) there is no correspondence between the target class **Writer** and a source class.

[4] https://www.dbpedia.org/.
[5] https://chat.openai.com/.

Figure 1 presents the recall, precision, and f1-score computed on the set of candidate instance descriptions. This figure shows the evolution of these measures for the different iterations. We have set the similarity threshold to 0.8. This figure shows that the recall increases with the number of iterations. We observe that the precision may decrease after three or four iterations in Figs. 1b and 1a (even if it is above the one achieved by the baseline as shown in Fig. 2). This means that there is a number of iterations after which the process is not effective. Identifying this number according to the characteristics of the sources will be the topic of future works. If no alignment between the target schema and the source is provided (as for the class Writer Fig. 1c), a higher number of iterations is required to achieve a good recall. This experiment shows that the iterative process enables the discovery of more candidate instance descriptions, especially when no class alignment is given.

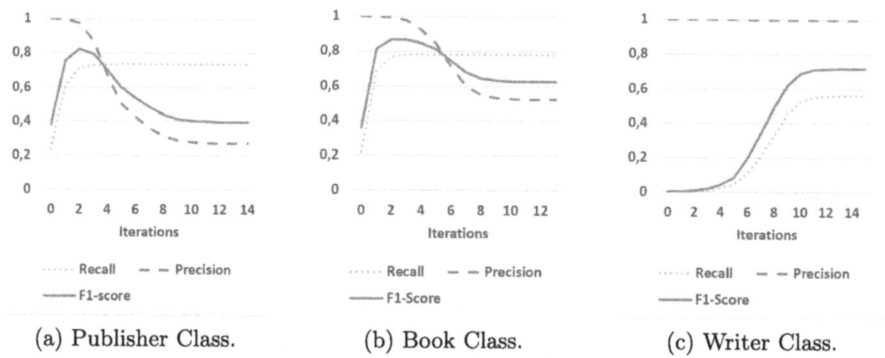

(a) Publisher Class.

(b) Book Class.

(c) Writer Class.

Fig. 1. Impact of the Number of Iterations on the Quality of the Result

Figure 2 shows the f1-score computed on the set of candidate instance descriptions for both the baseline approach introduced in Sect. 2, and our approach described in Sect. 4. The figure shows the impact of the similarity threshold on the f1-score. We can see on Fig. 2a that our approach achieves a higher f1-score than the baseline when the similarity threshold is above 0.7. Figure 2b shows that for each threshold, our approach has a higher f1-score. In the specific case of Writers (which has no equivalent class), fewer patterns are found with the baseline (with f1-score lower than 0.1) compared to our approach (with f1-score higher than 0.5).

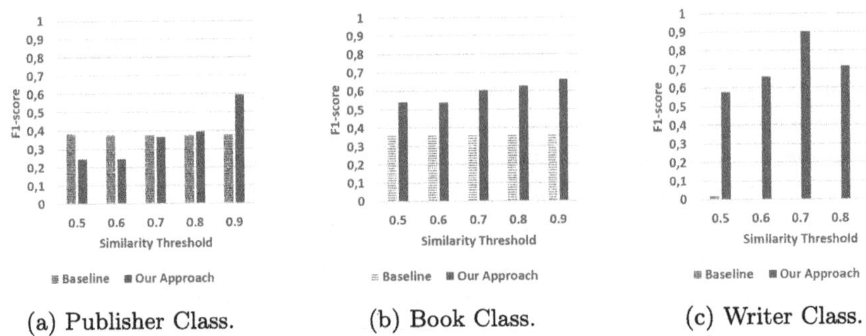

(a) Publisher Class. (b) Book Class. (c) Writer Class.

Fig. 2. Impact of the Similarity Threshold on the Target Schema Instantiation

6 Related Work

The generation of mappings has been the subject of several works, such as [2,6]. These approaches rely on the schema exported by the data sources, and such schema is not always provided for RDF datasets. In the RDF context, Sacramento et al. have addressed the problem of expressing a data source in the terms defined by an ontology [10], which consists in generating a mapping between a source and this ontology. This approach requires the alignment between the target and the source schemas, which is not always provided in our context.

A stream of works has addressed the problem of discovering relevant datasets in data lakes or equivalent data storage [1,5,7]. These approaches usually deal with structured data (e.g., CSV files) and use the notion of joinability and unionability. They rely on a predefined schema for both the target and the source, which are not always provided in our setting.

Some works address the problem of knowledge graph refinement, described in the survey paper proposed by Paulheim [8]. Some approaches address the problem of enriching a dataset by retrieving a type for untyped entities in an RDF data source [9]. These approaches deal with the inference of types based on the data. In our context, the types are only described in a target schema: there is no example of what an instance of the type is. Moreover, in our context, a typed entity in the source could also be a candidate to populate a class defined in the target schema.

To the best of our knowledge, the identification of candidate instances for a target schema when few or no alignment between the source and the target schema has not been addressed.

7 Conclusion and Future Work

In this paper, we have proposed a novel approach to identify existing patterns representing candidate instances for a target schema from RDF data sources. Our approach identifies candidate instance descriptions based on a semi-supervised learning algorithm that iteratively compares entities to the target classes by comparing them to both the class definition and the candidate instance patterns already identified for this class.

In our future works, we will focus on the automatic generation of the SPARQL queries to instantiate the target schema based on the identified candidate instance descriptions. We will also extend our approach by taking into account some constraints defined on the target schema so as to extract only the candidate instance descriptions for which the constraints are verified. These constraints could be defined using the W3C language SHACL[6].

References

1. Bogatu, A., Fernandes, A.A.A., Paton, N.W., Konstantinou, N.: Dataset discovery in data lakes. In: 2020 IEEE 36th International Conference on Data Engineering (ICDE), pp. 709–720. IEEE, Dallas, TX, USA, April 2020
2. Fagin, R., Haas, L.M., Hernández, M., Miller, R.J., Popa, L., Velegrakis, Y.: Clio: schema mapping creation and data exchange. In: Borgida, A.T., Chaudhri, V.K., Giorgini, P., Yu, E.S. (eds.) Conceptual Modeling: Foundations and Applications. LNCS, vol. 5600, pp. 198–236. Springer, Heidelberg (2009). https://doi.org/10.1007/978-3-642-02463-4_12
3. Fernández, J.D., Martínez-Prieto, M.A., de la Fuente Redondo, P., Gutierrez, C.: Characterising RDF data sets. J. Inf. Sci. **44**(2), 203–229 (2018)
4. Jaccard, P.: Distribution de la flore alpine dans le Bassin des Dranses et dans quelques régions voisines (1901)
5. Koutras, C., et al.: Valentine: evaluating matching techniques for dataset discovery. In: 37th IEEE International Conference on Data Engineering, ICDE 2021, Chania, Greece, 19–22 April 2021, pp. 468–479. IEEE (2021)
6. Mazilu, L., Paton, N.W., Fernandes, A.A., Koehler, M.: Schema mapping generation in the wild. Inf. Syst. **104**, 101904 (2022)
7. Miller, R.J.: Open data integration. Proc. VLDB Endowment **11**(12), 2130–2139 (2018)
8. Paulheim, H.: Knowledge graph refinement: a survey of approaches and evaluation methods. Semantic Web **8**(3), 489–508 (2017)
9. Paulheim, H., Bizer, C.: Type inference on noisy RDF data. In: Alani, H., et al. (eds.) ISWC 2013. LNCS, vol. 8218, pp. 510–525. Springer, Heidelberg (2013). https://doi.org/10.1007/978-3-642-41335-3_32
10. Sacramento, E.R., Vidal, V.M.P., de Macêdo, J.A.F., Lóscio, B.F., Lopes, F.L.R., Casanova, M.A.: Towards automatic generation of application ontologies. J. Inf. Data Manag. **1**(3), 535–550 (2010)
11. Yarowsky, D.: Unsupervised word sense disambiguation rivaling supervised methods. In: Proceedings of the 33rd Annual Meeting on Association for Computational Linguistics, pp. 189–196. Association for Computational Linguistics, Cambridge, Massachusetts (1995)

[6] https://www.w3.org/TR/shacl/.

Novelty-Driven Evolutionary Scriptless Testing

Lianne V. Hufkens[1(✉)], Tanja E. J. Vos[1,2], and Beatriz Marín[2]

[1] Open Universiteit, Valkenburgerweg 177, 6419 AT Heerlen, The Netherlands
lianne.hufkens@ou.nl

[2] Universitat Politècnica de València, València, Spain

Abstract. In recent years, scriptless Graphical User Interface (GUI) testing has been positioned as a complement to traditional testing techniques. Automated scriptless GUI testing approaches use Action Selection Rules (ASR) to generate on-the-fly test sequences when testing a software system. Currently, random is the standard selection approach in scriptless testing, provoking drawbacks in the testing process, such as test sequences that do not reflect the human strategies for testing, and being unable to deal with multistep tasks. This paper presents an alternate selection approach based on the use of a grammar to design the ASR and an Evolutionary Algorithm (EA) with Novelty Search (NS) to direct the evolution process. Preliminary testing shows that the ASRs do evolve in the standard EA process. Further research is needed to show the benefits of the additional NS for the testing process.

Keywords: Automated GUI testing · Scriptless testing · Novelty Search · Grammar-based testing · Evolutionary testing · Grammatical Evolution

1 Introduction

Within the field of automated GUI testing, scriptless testing is the approach to test software without a predefined test sequence. The underlying principle is simple: generate test sequences of (state, action)-pairs by starting up the System Under Test (SUT) in its initial state, and continuously select an action to bring the SUT into another state [27]. (See Fig. 1.) Selecting actions characterizes the fundamental challenge of intelligent systems, i.e. what to do next [34].

Currently, random selection is the standard approach in scriptless testing. This is because it is easy to implement, and also robust yet flexible in the face of change. It has already shown to be a valuable addition to the testing process in industrial contexts [2,5,6,25]. However, there are drawbacks. Random selection has no direction, cannot deal with multistep tasks, is not an accurate reflection of how users (humans) behave, and is highly unstructured.

We will contribute an alternate testing strategy, which is generic enough to not be hindered by changes to the SUT, and provides enough differentiation

J. Araújo et al. (Eds.): RCIS 2024, LNBIP 514, pp. 100–108, 2024.
https://doi.org/10.1007/978-3-031-59468-7_12

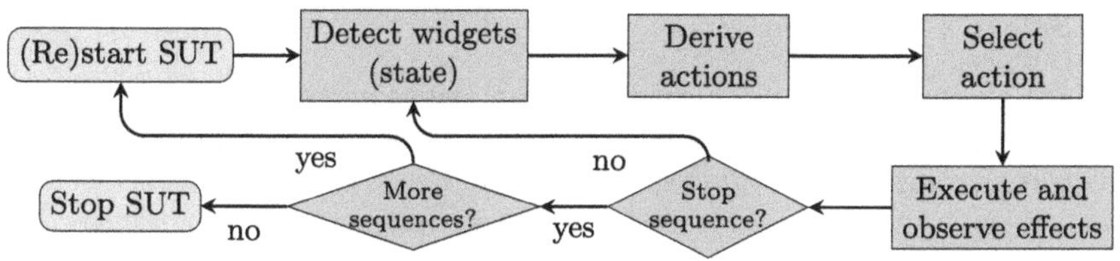

Fig. 1. Test sequence execution flow in scriptless testing.

between the available actions at any point in the process to steer the testing in a specified direction. To do that, we will use an Evolutionary Algorithm (EA) with Novelty Search (NS).

Using Evolutionary Algorithms (EA) on software engineering problems and reformulating them as optimization problems is known as Search-Based Software Engineering (SBSE) [12,15,20]. In this domain, we seek to automate the creation of new testing strategies using an Evolutionary Algorithm (EA). This is a system that adapts ('evolves') a pool of individuals (candidate solutions) to solve a problem by mimicking evolution in nature. After each round, only the most suitable individuals (as judged by the fitness function) are used to create the individuals that form the next generation. This process is repeated until termination. We additionally seek to modify EA with Novelty Search (NS) to gear the evolution process towards novel solutions.

This paper is structured as follows: Sect. 2 lists the components needed for a Novelty-driven Evolutionary Scriptless Testing environment, Sect. 3 describes the architecture, Sect. 4 covers the application of Novelty Search (NS), Sect. 5 lists related work, and finally Sect. 6 holds conclusions and future work.

2 Five Steps Towards Evolutionary Scriptless Testing

The goal of this research is to evolve new ASRs for scriptless testing using EA [14]. We will call this Novelty-driven Evolutionary Scriptless Testing (NEST).

For this we need five steps such that we can solve the problem:

1. a representation of the individuals (i.e. the ASRs)
2. a suitable EA variant
3. a set of manipulation operators that fits the EA variant
4. a suitable approach to assigning fitness
5. an architecture to join the EA and scriptless testing systems

2.1 Representation

The first part corresponds to the representation of the individuals , which allows the EA process to change and rearrange the genetic makeup of the individuals, as well as produce offspring needed for the next generation. To do that, we have developed a grammar with which testers can write test strategies. This grammar is described in more detail in [13]. A strategy written with this grammar is referred to as an Action Selection Rule (ASR).

2.2 Select EA Version

The most common EA variations use representations in the form of integer arrays (Genetic Algorithm, GA) [16], tree structures (Genetic Programming, GP) [28], or vectors (Evolution Strategy, ES) [4]. At first glance, the logical choice is the GP approach, as the ASRs can be expressed as trees. However, this brings the problem that the EA system itself does not adhere to the grammar's rules, and happily accepts invalid ASRs as long as the tree is correctly formed. For that reason, we have chosen to use a variation of EA known as Grammatical Evolution (GE). As the name indicates, this variety is explicitly meant for evolving individuals based on a context-free grammar.

Grammatical Evolution. Unlike other variations of EA, Grammatical Evolution (GE) separates the encoding of the genes ('genotype') and what the information represents ('phenotype'). GE's genotype are strings of numbers, the same form as in Genetic Algorithm (GA), while its phenotype matches Genetic Programming's (GP) tree structures. GE treats the numbers as instructions for which rules to choose from the grammar. As long as the grammar does not change, the same set of values always translates to the same individual (in this context, an ASR). Furthermore, the end result will always be a valid instance.

2.3 Manipulation Operators: Crossover and Mutation

Every round, the manipulation operators diversify the population and replace the individuals (i.e. ASRs) that did not make the cut with new individuals. Subsequently, they control how much the population changes with every generation. The two most common operators are *crossover* and *mutation*.

Crossover Two individuals act as 'parents' and reproduce. Their child is randomly assembled from copies of the components of their parents.

Mutation Randomly apply a change to one node of a chosen individual, in keeping with the constraints. It can possibly add or remove a set of children.

Typically, applications of *mutation* follow after *crossover* has concluded.

2.4 Assigning Fitness

Traditional fitness functions rely on translating the intended problem into a target to reach. It measures and judges the performance of individuals (i.e. the ASRs) [12] to assign a score that reflects how 'fit' they are to solve the problem. The individuals with better fitness scores will pass their good 'genes' on to the next generation, as a basis for further improvement. However, it is possible for the goal to be uncertain, hard to measure, or simply (accurately) translatable into a fitness function. Defining the right fitness function is not an easy task; the goal of testing is to find faults, but not finding any is not necessarily a proof that the testing process was adequate. Testing harbours a paradox [3,23]: one never knows beforehand *if* there are bugs to be found at all, and if so, how many there are, where they are, or their severity. Having any of this information beforehand would make testing easier and more efficient, but at the same time take away the reason to test at all. After all, there is no point in expending effort when the answer is already known. Thus, every application of SBSE to testing grapples with the same question: how to set the goal (find bugs) when it does not know where they are or even if there are any? Therefore, there is no known goal for a fitness function.

2.5 Proposed Architecture

The intended setup is shown in Fig. 2. Left (red) is the EA system, with its internal loop to evolve the population of individuals (candidate strategies in this case). Top right (blue-purple) shows the scriptless testing system, and SUT (light orange) represents the software system used to evaluate the performance of the candidate strategies. The bottom right (green) part is a representation of the databases and logs that will collect and record the data (metrics) from the testing tool, then enter it back into the EA system.

3 Implemented Architecture

We have implemented the setup of the proposed architecture by using PonyGE2 [10] for the EA-part and TESTAR [34] as the scriptless testing tool. Preliminary test runs show that the ASRs are indeed capable of evolving in a standard GE system with goal-based fitness. The SUT used is the Parabank demo website [24], which mimics a banking application. It provides fictional bank services, including search options for financial transactions like bill payments. The next step is to implement NS in place of the goal-based fitness. The final architecture is depicted in Fig. 2. We will describe each of the components in the following subsections.

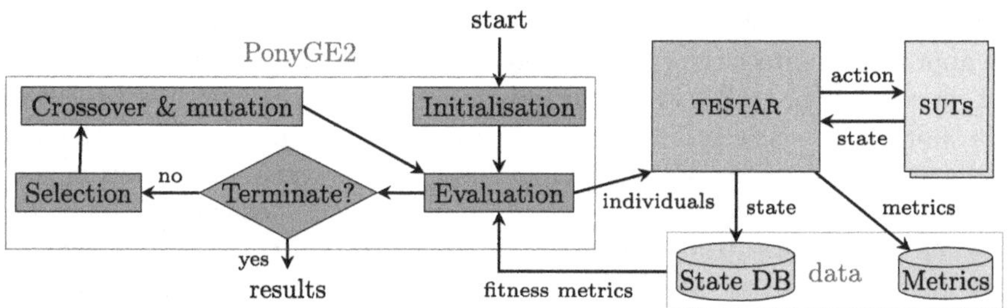

Fig. 2. The proposed evolutionary system (Color figure online)

3.1 PonyGE2

As GE system, we chose PonyGE2 [10]. It is a compact, highly customizable, modular program written in Python, that offers many settings (like options for both single- and multi-objective search), and predefined components, such as operators and ready-made templates for new fitness functions.

3.2 TESTAR

Our chosen scriptless system is TESTAR [30,31]. A test run with TESTAR works as follows (see Fig. 1): start the SUT, detect the current state (i.e. all available control elements (widgets)) of the GUI, derive all possible actions for these widgets, select one action using the current ASR, execute that action, and observe the effects. This flow is repeated until a stop criterion is met, e.g. when a previously defined sequence length has been reached or a crash has been detected.

3.3 State Model

During testing, TESTAR can track the actions and states visited in a state model. Internally, it uses an OrientDB graph database to store the model [26]. Nodes are states, actions are recorded as edges, and as testing progresses, the SUT's internal structure is captured in the shape of the graph. It also records the information of the widgets present in the states. This makes the state model a suitable source of information to record, and later extract, the paths ASRs have taken through the same SUT(s). This is necessary to calculate how similar or differently the individuals of the population have behaved during the evaluation phase of EA (see Fig. 2). And this, in turn, is used to calculate how 'novel' every individual is in its testing approach.

4 Applying Novelty Search for Testing

Novelty Search (NS) steps away from goal-based search by providing an alternative approach to direct the evolution process [18,19]. It ignores the concept of a goal entirely and instead rewards the individuals showing entirely new behaviour. This approach incentivizes creativity, assigning better fitness values to the more unique individuals. After all the common 'easy' approaches have been exhausted, (NS) forces the evolution to move on to more complex behaviours. While this does not directly lead to better testing, it should produce a variety of different takes on testing, among which there should be better strategies.

At first glance, the concept of NS appears to be related to Curiosity Learning (CL), most commonly used in Reinforcement Learning (RL), an algorithm in the field of Machine Learning. While both concepts do share similarities in that they encourage exploration, they are still very different mechanisms. RL sees agents explore and map out an environment without any prior knowledge. The system only communicates with agents for reaching goal states (positive reward), or dead ends (negative reward). CL is an approach to encourage agents to explore unknown areas and seek out new things, but does not direct behaviour.

For NS to work in our setup, it is crucial to define what 'novel' means. This determines in what way the best of the individuals must be different from the rest, and what metric is needed to differentiate between behaviours. The starting point is clear: finding new bugs will only happen if the process is constantly trying new routes. Visiting the same part of the SUT again and again serves no purpose, and, similarly, there could be too few unique states or actions available to limit the process to visit only new parts of the SUT.

Changing the way states are entered can be the difference between finding a bug and missing it altogether. Additionally, the abstraction level determines what counts as a unique state, and is thus an important factor influencing the fitness values [34]. An equilibrium between the necessary expressiveness of the states and the computational complexity [21] should be found. These paths are series of actions, just like string values are series of characters. This means that we can use string metrics (like edit distance) to compare and analyse the paths.

5 Related Work

EA in testing has mainly been focused on the generation of input data. However, there are examples where EA (and alternatives) have been used to direct the testing process. [17] uses EA to evolve test cases for GUI testing. Similarly, [8] makes use of Ant Colony Optimization (an algorithm from the same family as EA), later enhanced by Q-learning, to find optimum testing paths in SUTs. [22] generates test sequences using a variety of different techniques, among with were three versions of Multi-Objective Evolutionary Algorithm (MOEA).

For TESTAR, attempts to create suitable fitness functions have been made. In [9,11,32] the fitness function was "the number of different (abstract) states visited during testing". Additionally, in [29] other fitness functions have been defined that focus on introducing "new actions" more often, achieving high "state model coverage" or "code coverage". For all of them, how much the fitness functions influenced test effectiveness was inconclusive, precisely because the goal of testing is hard to capture with a fitness function. Most current applications of Novelty Search focus on generating test sequences or use cases. This is not the same as our intention to evolve generic strategies (ASRs), which are meant to dynamically select actions in real-time. [7] used GA with NS to identify user cases. [1] used a hybrid that combines GA and hill climbing with NS to generate test suites (consisting of test cases) for Android applications. [33] combines GP with NS for automatically detecting and repairing bugs in code.

6 Conclusions and Future Work

We look for the feasibility to add Novelty-driven EA to the grammar-based testing we already have, and aim to demonstrate that it is possible to improve the testing process. The resultant ASR(s) are suitable for generic testing, which we plan to use in order to encourage the evolution by using a pool of different SUTs, consisting of websites, desktop applications, and android applications. For future work, we also would like to establish if grammar-based ASRs are indeed human-like in behaviour.

References

1. Amalfitano, D., Amatucci, N., Fasolino, A.R., Tramontana, P.: Agrippin: a novel search based testing technique for android applications. In: 3rd International Workshop on Software Development Lifecycle for Mobile, pp. 5-12. DeMobile 2015, ACM (2015)
2. Bauersfeld, S., Vos, T.E., Condori-Fernández, N., Bagnato, A., Brosse, E.: Evaluating the testar tool in an industrial case study. In: 8th ACM/IEEE International Symposium on Empirical Software Engineering and Measurement, pp. 1–9 (2014)
3. Bertolino, A.: Software testing research: achievements, challenges, dreams. In: Future of Software Engineering (FOSE'07), pp. 85–103. IEEE (2007)
4. Beyer, H.G., Schwefel, H.P.: Evolution strategies - a comprehensive introduction. Nat. Comput. **1**(1), 3–52 (2002)
5. Bons, A., Marín, B., Aho, P., Vos, T.E.: Scripted and scriptless gui testing for web applications: an industrial case. Inf. Softw. Technol. **158**, 107172 (2023)
6. Chahim, H., Duran, M., Vos, T.E.J., Aho, P., Condori Fernandez, N.: Scriptless Testing at the GUI Level in an Industrial Setting. In: Dalpiaz, F., Zdravkovic, J., Loucopoulos, P. (eds.) Research Challenges in Information Science: 14th International Conference, RCIS 2020, Limassol, Cyprus, September 23–25, 2020, Proceedings, pp. 267–284. Springer International Publishing, Cham (2020). https://doi.org/10.1007/978-3-030-50316-1_16
7. DeVries, B., Trefftz, C.: A novelty search and metamorphic testing approach to automatic test generation. In: 2021 IEEE/ACM 14th International Workshop on Search-Based Software Testing (SBST), pp. 8–11 (2021)
8. Dorigo, M., Blum, C.: Ant colony optimization theory: a survey. Theoret. Comput. Sci. **344**(2), 243–278 (2005)
9. Esparcia-Alcázar, A., Almenar, F., Vos, T.E.J., Rueda, U.: Using genetic programming to evolve action selection rules in traversal-based automated software testing: results obtained with the testar tool. Memetic Comput. **10**(3), 257–265 (2018)
10. Fenton, M., McDermott, J., Fagan, D., Forstenlechner, S., Hemberg, E., O'Neill, M.: Ponyge2: grammatical evolution in python. In: Proceedings of the Genetic and Evolutionary Computation Conference Companion. GECCO '17, ACM (Jul 2017)
11. de Groot, M.: Using evolutionary computing to improve black box monkey testing on a Graphical User Interface. Master's thesis, Open Universiteit, Heerlen, Netherlands (Apr 2018)
12. Harman, M., Jones, B.F.: Search-based software engineering. Inform. Softw. Technol.**43**, 833–839 (2001)
13. Hufkens, L.V.: Grammar-based action selection rules for scriptless testing. In: To be published in 5th ACM/IEEE International Conference on Automation of Software Test (AST). ACM (2024)
14. Hufkens, L.V.: Evolutionary scriptless testing. In: Guizzardi, R., Ralyté, J., Franch, X. (eds.) Research Challenges in Information Science: 16th International Conference, RCIS 2022, Barcelona, Spain, May 17–20, 2022, Proceedings, pp. 779–785. Springer International Publishing, Cham (2022). https://doi.org/10.1007/978-3-031-05760-1_55

15. Khari, M., Kumar, P.: An extensive evaluation of search-based software testing: a review. Soft. Comput. **23**(6), 1933–1946 (2019)
16. Lambora, A., Gupta, K., Chopra, K.: Genetic algorithm- a literature review. In: 2019 International Conference on Machine Learning, Big Data, Cloud and Parallel Computing (COMITCon), pp. 380–384 (2019)
17. Latiu, G.I., Creţ, O., Văcariu, L.: Evoguitest - a graphical user interface testing framework based on evolutionary algorithms. In: 5th International Joint Conference on Computational Intelligence - Volume 1: ECTA. IJCCI 2013, pp. 75–82. SciTePress, INSTICC (2013)
18. Lehman, J., Stanley, K.O.: Efficiently evolving programs through the search for novelty. In: Proceedings of the 12th Annual Conference on Genetic and Evolutionary Computation, pp. 837–844. ACM, Portland Oregon USA (Jul 2010)
19. Lehman, J., Stanley, K.O.: Novelty search and the problem with objectives. In: Riolo, R., Vladislavleva, E., Moore, J.H. (eds.) Genetic programming theory and practice IX, pp. 37–56. Springer, New York, New York, NY (2011)
20. McMinn, P.: Search-based software testing: past, present and future. In: 2011 IEEE Fourth International Conference on Software Testing, Verification and Validation Workshops, pp. 153–163 (March 2011)
21. Meinke, K., Walkinshaw, N.: Model-based testing and model inference. In: Margaria, T., Steffen, B. (eds.) Leveraging Applications of Formal Methods, Verification and Validation. Technologies for Mastering Change, pp. 440–443. Springer Berlin Heidelberg, Berlin, Heidelberg (2012). https://doi.org/10.1007/978-3-642-34026-0_32
22. Menninghaus, M., Wilke, F., Schleutker, J.P., Pulvermüller, E.: Search based gui test generation in java - comparing code-based and efg-based optimization goals. In: Proceedings of the 12th International Conference on Evaluation of Novel Approaches to Software Engineering. vol. 1, pp. 179–186. INSTICC, SciTePress (2017)
23. Papanikolaou, K.: Software testing: a never-ending adventure. https://ginbits.com/software-testing-a-never-ending-adventure/. Accessed 1 Feb 2024
24. Parasoft: Parabank demo application (2017, 2022). https://github.com/parasoft/parabank/. Accessed 2 Feb 2022
25. Ricós, F.P., Aho, P., Vos, T., Boigues, I.T., Blasco, E.C., Martínez, H.M.: Deploying TESTAR to enable remote testing in an industrial CI pipeline: a case-based evaluation. In: Margaria, T., Steffen, B. (eds.) Leveraging Applications of Formal Methods, Verification and Validation: Verification Principles: 9th International Symposium on Leveraging Applications of Formal Methods, ISoLA 2020, Rhodes, Greece, October 20–30, 2020, Proceedings, Part I, pp. 543–557. Springer International Publishing, Cham (2020). https://doi.org/10.1007/978-3-030-61362-4_31
26. Ricós, F.P., Neeft, R., Marín, B., Vos, T.E.J., Aho, P.: Using GUI change detection for delta testing. In: Nurcan, S., Opdahl, A.L., Mouratidis, H., Tsohou, A. (eds.) Research Challenges in Information Science: Information Science and the Connected World: 17th International Conference, RCIS 2023, Corfu, Greece, May 23–26, 2023, Proceedings, pp. 509–517. Springer Nature Switzerland, Cham (2023). https://doi.org/10.1007/978-3-031-33080-3_32
27. Pastor Ricós, F., Slomp, A., Marín, B., Aho, P., Vos, T.E.: Distributed state model inference for scriptless GUI testing. J. Syst. Softw. **200**, 111645 (2023)
28. Poli, R., Langdon, W.B., McPhee, N.F., Koza, J.R.: A field guide to genetic programming. Lulu Press], [Morrisville, NC (2008)
29. Stoyanov, N.: Strategy based genetic algorithms approach in automated GUI testing. Master's thesis, TU/e, Eindhoven, Netherlands (Sep 2020)

30. TESTAR: Testar on github. https://github.com/TESTARtool/TESTAR_dev
31. TESTAR: Testar project download page. https://testar.org/download/
32. Theuws, G.: Random action selection vs genetic programming: a case study in TESTAR. Master's thesis, Open Universiteit, Heerlen, Netherlands (Feb 2020)
33. Villanueva, O.M., Trujillo, L., Hernandez, D.E.: Novelty search for automatic bug repair. In: Proceedings of the 2020 Genetic and Evolutionary Computation Conference, pp. 1021-1028. GECCO '20, ACM, New York, NY, USA (2020)
34. Vos, T.E.J., Aho, P., Ricos, F.P., Rodriguez-Valdes, O., Mulders, A.: Testar - scriptless testing through graphical user interface. Softw. Test. Verif. Reliab. **31**(3), e1771 (2021)

Doctoral Consortium Papers

Strengthening Cloud Applications: A Deep Dive into Kill Chain Identification, Scoring, and Automatic Penetration Testing

Stefano Simonetto(✉)

Department of Pervasive Systems, University of Twente, Enschede, The Netherlands
s.simonetto@utwente.nl

Abstract. The need to anticipate and defend against potential threats is paramount in cybersecurity. This study addresses two fundamental questions: what attacks can be performed against my system, and how can these attacks be thwarted?

Addressing the first question, this work introduces an innovative method for generating executable attack programs, showcasing the practicality of potential breach scenarios. This approach not only establishes the theoretical vulnerability of a system but also underscores its susceptibility to exploitation.

To respond to the second question, the proposed approach explores a range of mechanisms to counter and thwart the exposed attack strategies. The aim is to use robust and adaptive defensive strategies, leveraging insights from the demonstrated attack programs. These mechanisms encompass proactive measures, such as automatic penetration testing and behavior analysis, and reactive approaches, such as rapid patch deployment and vulnerability prioritization. The resilience of systems against potential breaches can be enhanced by intertwining attack pathways with comprehensive countermeasures, thereby disrupting the adversary's kill chains. This study aims to contribute to the containerized application security deployed in different environments, like the Cloud, Edge, 5G, Internet of Things (IoT), and Industrial IoT (IIoT), by taking these scenarios as a case study.

This research contributes to the evolution of cyber threat analysis through a Design Science Research (DSR) approach, focusing on developing and validating artifacts, tools, and frameworks. Defenders can anticipate, combat, and ultimately mitigate emerging threats in an increasingly complex digital environment by creating tangible attack programs and formulating effective thwarting mechanisms.

Keywords: Vulnerability · Prioritization · Penetration testing · Kill chain

1 Introduction

Cloud application vulnerabilities devastate our digital society, threatening privacy, finances, and critical infrastructure. Businesses must recognize this threat

J. Araújo et al. (Eds.): RCIS 2024, LNBIP 514, pp. 111–120, 2024.
https://doi.org/10.1007/978-3-031-59468-7_13

and safeguard their company from cloud vulnerabilities. A 2021 study by IBM suggests that data breaches caused by cloud security vulnerabilities cost companies an average of $4.8 million to recover [14].

In the past few years, the research community has proposed sophisticated approaches and techniques to enhance automated security testing and promptly identify vulnerabilities before malicious attackers can exploit them.

As a result, many tools are available today to detect vulnerabilities effectively. However, most organizations do not know how to deal with hundreds of vulnerabilities because these tools are prone to produce false positives. The usual behavior is to patch them based on the produced vulnerability's score.

As if the problem wasn't complicated enough, the container orchestration scenario makes the situation more challenging due to the rapid and continuous deployment of new containers and pods in such environments.

In the real world, attackers combine various vulnerabilities to breach systems effectively; thus, analyzing vulnerabilities in combination with each other represents a fundamental step to obtain a realistic "big picture" of their implications.

Furthermore, most defensive solutions are reactive, like intrusion detection systems, system calls monitoring, etc. Even if these techniques are well-established, they are affected by scalability problems and are not meant to prevent an attack from happening. The remainder of this paper is organized as follows. Section 2 gives some background information and a taxonomy of fundamental principles that guided the present work. Section 3 presents the research design. Finally, Sect. 4 presents the conclusions, and Sect. 5 outlines the principles of open science, which aim to democratize knowledge, increase transparency and collaboration, and enhance scientific research's quality and impact.

2 Background and Taxonomy

This section gives background information and a taxonomy of fundamental principles to understand the research questions and objectives.

2.1 Architecture

In the ever-growing landscape of software development and applications, technologies such as Docker [1], Kubernetes [5], and cloud providers are emerging as transformational forces. They represent a fundamental shift in how applications are built, managed, and scaled in today's computing environment. Docker, introduced in 2013, revolutionized containerization by packaging applications and their reliance on units called containers. These lightweight, portable containers enable developers to create consistent, isolated environments that run smoothly on any platform, from local development machines to production servers. Kubernetes, often shortened to K8s, emerged in 2014 as an open-source container orchestration platform developed by Google. It supports containerization, which can provide a more efficient and effective way to manage, scale, and deploy containerized applications. K8s abstracts the underlying infrastructure, letting developers focus on defining desired application states, and the system handles the complex task of managing policies and containers across clusters of machines as depicted in Fig. 1. Cloud providers such as Amazon Web Services, Microsoft Azure, Google Cloud Platform, etc., have greatly influenced the way infrastructure and IT resources are provisioned, managed, and utilized [15]. These providers offer a vast array of services, including computing power, storage, networking, and databases, accessible over the Internet on a pay-as-you-go basis. Businesses can leverage these technologies to scale resources on demand, reduce hardware costs, and achieve flexibility and agility in their operations.

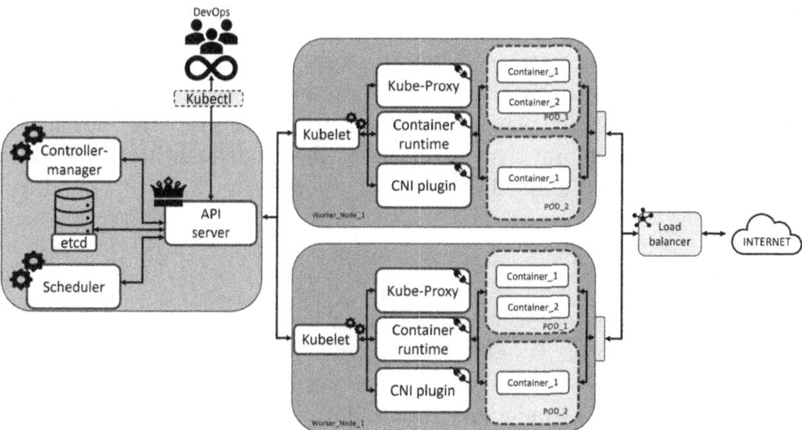

Fig. 1. Kubernetes simple scenario

2.2 Vulnerabilities and Frameworks

In the rapidly evolving landscape of cybersecurity, protecting digital assets from potential threats and attacks is a constant challenge. Understanding and addressing vulnerabilities is essential to strengthen systems and networks effectively. Three crucial elements that form the foundation of vulnerability management are Common Vulnerabilities and Exposures (CVEs) [24], Common Weakness Enumeration (CWEs) [23], and misconfigurations [17]. Furthermore, the Common Vulnerability Scoring System (CVSS) [19] serves as a standardized method for assessing and quantifying the severity of identified vulnerabilities.

Understanding adversary behavior is crucial in cybersecurity. Two approaches exist for organizing information about adversarial actions: Common Attack Pattern Enumeration and Classification (CAPEC) and Adversarial Tactics Techniques & Common Knowledge (ATT&CK). Each is tailored for distinct use cases. MITRE ATT&CK [3] is a globally accessible knowledge base of adversary tactics and techniques based on real-world observations. While CVE, CWE, and CVSS provide essential details about individual vulnerabilities and their severity, the ATT&CK framework comprehensively explains how attackers might exploit them within various cyber attack stages. The ATT&CK knowledge base creates particular threat models and methodologies in cybersecurity products and service communities.

On the other hand, CAPEC [22] focuses on application security and delineates the typical characteristics and methods attackers use to take advantage of recognized weaknesses in cyber-enabled capabilities (e.g., SQL Injection, XSS).

2.3 Effective Vulnerability Analysis

The main challenge for an effective vulnerability analysis and prioritization strategy is considering multiple vulnerabilities and their combined capabilities. In real-world settings, attackers put together ("chain") multiple vulnerabilities to successfully compromise systems. Thus, ranking vulnerabilities individually is insufficient and unrealistic. Hence, certain companies rely on penetration testers to thoroughly assess their security measures to understand the potential kill chains [13]. This concept involves structured phases delineating the attacker's advancement towards accomplishing goals.

2.4 Cloud, Edge and IoT

As highlighted in the work by Koziolek, H. et al. [16], the dominant approach to software deployment is rapidly shifting towards containerization. Simultaneously, there is a growing fascination with employing container orchestration frameworks, extending beyond conventional data centers to encompass resource-limited hardware like Internet-of-Things devices, edge gateways, and more.

The ongoing effort to extend container orchestration to the edge represents a significant adaptation of containerization technology [8]. While container orchestrators were initially designed for managing cloud-based applications, they are now being applied to edge computing environments. This transition is driven by the need to efficiently manage and deploy containerized applications on resource-constrained devices like IoT devices and Edge gateways. Containers make building, deploying, and maintaining IoT applications easier, even when IoT devices have limited resources to support operating systems. This approach enables the deployment of containerized applications closer to where data is generated, reducing latency and enhancing real-time processing capabilities.

As shown in Fig. 2, the containerization, orchestration, and the different use cases add layers to the bare code that needs to run. Thus, in turn, it can bring new specific vulnerabilities and misconfigurations, making the overall scenario more complex.

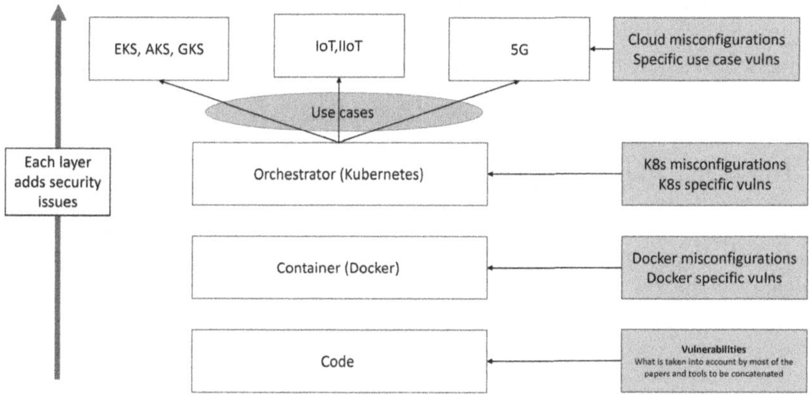

Fig. 2. Complete scenario

3 Research Design

This study aims to encourage collaboration and the generation of more resilient solutions for addressing security issues in containerized applications. This section outlines the research objectives and provides an overview of the anticipated outcomes within the research plan.

3.1 Research Objective

The research goals can be summarized into three main points.

1. Associating vulnerabilities with the ATT&CK framework while emphasizing kill chains: when considering vulnerability concatenation, the relationship between vulnerabilities and the ATT&CK framework gains depth. This concept illustrates how combining vulnerabilities, CVEs, CWEs, and misconfigurations shapes the ATT&CK matrix. Security becomes more proactive and holistic by recognizing and addressing vulnerabilities in conjunction.
2. Comprehensive security landscape understanding: gain a comprehensive security perspective by analyzing vulnerabilities within the context of the kill chain. Comprehend how these vulnerabilities align, facilitating their prioritization and aggregation for achieving maximal potential impact.
3. Automating penetration testing within container orchestration settings: conduct a study on vulnerability discovery-fix automation processes. Investigate the interplay between vulnerability discovery automation, the ATT&CK framework, and proactive defenses. Utilize insights to automate penetration testing, identifying and addressing vulnerability kill-chains within the container applications environment.

The first objective will provide understanding and reasoning about kill chains and how they can be mapped to the MITRE framework. Once the kill chain has been recognized, the next step is to prioritize according to the most dangerous kill chain. Finally, the third objective aims to automatically produce kill chains, knowing what a kill chain looks like and how it can be produced in the most lethal way given a certain scenario.

3.2 Research Planning

A series of chronological sub-research questions have been formulated to achieve the research objectives. These sub-research questions will be addressed by the conclusion of the Ph.D. program. An overview of the research flow and expected output is provided in Fig. 3.

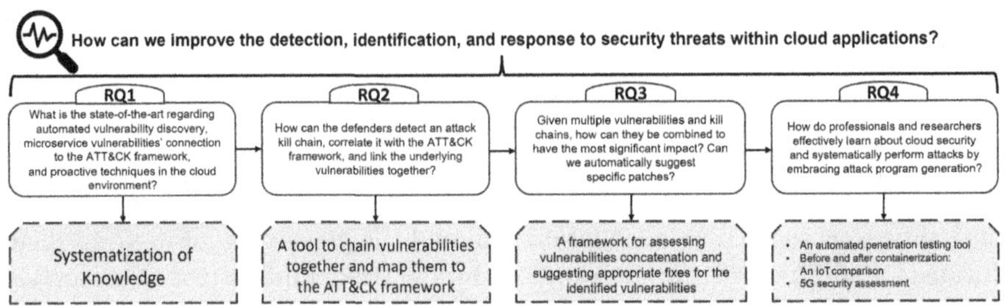

Fig. 3. Schematic research plan

3.3 Research Questions

The research question, which serves as the central query encapsulating the entire research concept and guiding the investigation, can be characterized as follows:

- How can we improve the detection, identification, and response to security threats within cloud applications?

To enhance clarity, the main question should be deconstructed into four sub-questions, each requiring answers about 'What' 'Why' and 'How. The research is still in the early stages. Still, the foundations have been laid, and tools will be developed according to the best technology when tackling the particular problem. The author acknowledges that the 'How' can be better defined, but the DSR framework [12] will be used for understanding, executing, and evaluating.

RQ1. What is the state-of-the-art regarding automated vulnerability discovery, microservice vulnerabilities' connection to the ATT&CK framework, and proactive techniques in the cloud environment?

What: This question aims to provide a comprehensive and structured review of the existing literature, research, and knowledge related to the attacks on cloud architecture.

Why: The need for Systematization of Knowledge (SoK) in the literature is paramount, and Usenix's recent invitations to provide SoKs have proven highly valuable in assisting the security community in clarifying and contextualizing complex research problems. In this study, underexplored research areas will be identified, methodologies and tools will be thoroughly evaluated, historical context will be provided, existing approaches will be critically assessed regarding strengths and limitations, and potential future research directions with suggested improvements will be outlined.

How: The methodology involves formulating search queries that correspond to the research questions, followed by an extensive review of papers and grey literature to pinpoint deficiencies in the existing literature. The research question aims to conduct a comprehensive review of the state-of-the-art by systematically summarizing and analyzing existing literature to gain insights into automatic vulnerability discovery, exploitation, and patching, the relationship between microservices and the ATT&CK framework, and the use of active and proactive techniques in cloud environments.

RQ2. How can an attack kill chain be identified, matched to the ATT&CK framework, and the underline vulnerabilities be concatenated?

What: This question aims to understand how an attack kill chain can be identified and mapped to the ATT&CK framework.

Why: To the author's best knowledge, the literature does not focus on this mapping between CVEs, CWEs, and misconfiguration regarding the cloud environment. Similarly, only a few articles like [9] and [11] are trying to relate the vulnerabilities generated by a real-world scenario to the ATT&CK framework and automatically retrieve the CVEs, CWEs, and misconfigurations that enable a particular technique/tactic to be performed by the attacker. Minna, F. et al. [18] present a Sok on run-time security for cloud microservices, emphasizing that there is room to improve tools for microservices by adding functionality to correlate exploitation steps to MITRE ATT&CK tactics and CVEs that might be exploited. Furthermore, there are no tools that can automatically find an attack kill chain in this scenario.

How: In the complex environment of Cloud Native applications, carrying out vulnerability management is challenging. Vulnerability discovery has been heavily studied in the past years, to the point where tools like fuzzers and scanners are becoming arguably too good, and we are finding more vulnerabilities than we can properly fix, leading to alert fatigue [21]. This problem occurs when cybersecurity professionals become desensitized after dealing with overwhelming alerts. To fill this gap, this research will focus on designing a tool that can be integrated into the CI/CD pipeline to help developers and security teams check for vulnerabilities that enable a specific tactic in the ATT&CK framework.

RQ3. Given multiple vulnerabilities and kill chains, how can they be combined to have the biggest impact? Can we automatically suggest specific patches?

What: This question aims to develop an automated vulnerability assessment system that utilizes real-world vulnerabilities and kill chains to propose effective, context-aware fixes.

Why: A primary gap in the literature is the limited availability of valuable datasets for vulnerability analysis. Emphasizing the concatenation is essential because assessing vulnerabilities solely based on their individual severity is both inadequate and impractical. Once the kill chains are identified, the next open problem in the literature is to propose context-aware patches automatically.

How: Creating a framework for the novel score-based rule for assessing vulnerabilities in real-world scenarios, particularly focusing on the vulnerabilities concatenation and the blast radii. Introducing a novel technique enhances the assessment of vulnerabilities in real-world scenarios, focusing on understanding how vulnerabilities concatenate. It intends to develop a score-based rule focusing on blast radii by employing a metric such as the CVSS environmental score to evaluate vulnerabilities. Subsequently, the research aims to leverage generative AI or other approaches to suggest appropriate fixes for the identified vulnerabilities.

RQ4. How do professionals and researchers effectively learn about cloud security and systematically perform attacks in this domain?

What: This research question aims to create an automated penetration testing tool that unifies established vulnerable scenarios, incorporates potential automated pen testing solutions, and integrates prevalent kill chains. Furthermore, it sheds light on the interplay between IoT/IIoT and 5G regarding their security implications within containerization.

Why: Due to the lack of repositories where researchers and security teams can experiment with security practices in the cloud application environments, efforts will be made to establish a unified repository for vulnerable scenarios to streamline the availability and access to useful and on-purpose vulnerable scenarios. A significant emphasis within this work will be on achieving the effective compilation of kill chains to provide an exhaustive comprehension of security threats because only [7] is trying to model the chaining process in such a scenario.

It is crucial to prioritize security and portability integration in IoT/IIoT scenarios, but formalization in this field is weak. Furthermore, the security aspect of container orchestration in the intricate setting of 5G networks will be examined, as this specific concern has not been covered in existing literature.

How: This research aims to create an automated penetration testing service by improving consolidated known vulnerable scenarios like [2], exploring solutions for automating pen testing, and combining effective kill chains. As highlighted in [10], today's attacks are not fine-tuned to microservices architecture, so there is room for improvement. The primary objective is to gather various Kubernetes security tools, such as Pirates [4], Kube-hunter [6], Kubeaudit [20], and others, and integrate their functionalities.

Comparing IoT/IIoT security before and after containerization and validating the results using a formal method such as fault trees.

Leveraging the previously mentioned tools to highlight the issues introduced by the container orchestration in 5G (RAN and core) and spot new attack paths that are introduced.

4 Conclusions

In conclusion, this research will dig into the vulnerability concatenation realm, aligning it with the ATT&CK framework and emphasizing the crucial aspect of kill chains. The concept of vulnerability concatenation, encompassing vulnerabilities, CVEs, CWEs, and misconfigurations, provides a deeper understanding of the intricate nature of cyber threats. Doing so equips developers and security teams to proactively guard against attacks across the entire kill chain, recognizing and addressing the compounding impact of exploiting multiple weaknesses together. Moreover, this research underscores the importance of gaining a comprehensive security perspective by analyzing vulnerabilities in the context of the kill chain. This approach aids in prioritizing and aggregating vulnerabilities based on their alignment with specific phases, thereby maximizing their potential impact. Such an approach is particularly beneficial in the IoT/IIoT scenario.

5 Open Science Principles

In line with open science principles, this research is committed to promoting accessibility and transparency. The study will openly share the results and tools developed with the scientific community and the public, promoting a collaborative and inclusive approach to cybersecurity research. The primary platform for sharing code and outcomes will be GitHub, whereas the research papers will be open-access. This will allow others to build on the findings and contribute to the advancement of the field.

Acknowledgment. The author likes to thank his supervisors, H.G. Peter Bosch, Paul J.M. Havinga, and Willem Jonker, for their contributions to this line of work.

References

1. Docker website. https://www.docker.com/. Accessed 21 Mar 2024
2. Kubernetes goat. https://github.com/madhuakula/kubernetes-goat. Accessed 21 Mar 2024
3. Matrix - Enterprise | MITRE ATT&CK. https://attack.mitre.org/matrices/enterprise/containers/. Accessed 21 Mar 2024
4. Peirates. https://github.com/inguardians/peirates. Accessed 21 Mar 2024

5. Production-grade container orchestration. https://kubernetes.io/. Accessed 21 Mar 2024
6. Aquasecurity: Kube-hunter (2023). https://github.com/aquasecurity/kube-hunter
7. Blaise, A., Rebecchi, F.: Stay at the helm: secure kubernetes deployments via graph generation and attack reconstruction. In: 2022 IEEE 15th International Conference on Cloud Computing (CLOUD), pp. 59–69 (2022). https://doi.org/10.1109/CLOUD55607.2022.00022
8. Goethals, T., De Turck, F., Volckaert, B.: Fledge: kubernetes compatible container orchestration on low-resource edge devices. In: Hsu, C.H., Kallel, S., Lan, K.C., Zheng, Z. (eds.) IOV 2019. LNCS, vol. 11894, pp. 174–189. Springer, Cham (2019). https://doi.org/10.1007/978-3-030-38651-1_16
9. Grigorescu, O., Nica, A., Dascalu, M., Rughinis, R.: CVE2ATT&CK: BERT-based mapping of CVEs to MITRE ATT&CK techniques. Algorithms **15**(9), 314 (2022)
10. Gupta, C., van Ede, T., Continella, A.: Honeykube: designing and deploying a microservices-based web honeypot. In: SecWeb 2023 (2023)
11. Hemberg, E., et al.: Linking threat tactics, techniques, and patterns with defensive weaknesses, vulnerabilities and affected platform configurations for cyber hunting. arXiv preprint arXiv:2010.00533 (2020)
12. Hevner, A.R., March, S.T., Park, J., Ram, S.: Design science in information systems research. MIS Q. 75–105 (2004)
13. Hutchins, E.M., Cloppert, M.J., Amin, R.M., et al.: Intelligence-driven computer network defense informed by analysis of adversary campaigns and intrusion kill chains. In: Leading Issues in Information Warfare & Security Research, vol. 1, no. 1, p. 80 (2011)
14. IBM Security: Cost of a data breach - a view from the cloud 2021 (2021). https://www.ibm.com/downloads/cas/JDALZGKJ
15. Kaushik, P., Rao, A.M., Singh, D.P., Vashisht, S., Gupta, S.: Cloud computing and comparison based on service and performance between amazon AWS, Microsoft Azure, and google cloud. In: 2021 International Conference on Technological Advancements and Innovations (ICTAI), pp. 268–273. IEEE (2021)
16. Koziolek, H., Eskandani, N.: Lightweight kubernetes distributions: a performance comparison of MicroK8s, k3s, k0s, and Microshift. In: Proceedings of the 2023 ACM/SPEC International Conference on Performance Engineering (2023)
17. Loureiro, S.: Security misconfigurations and how to prevent them. Netw. Secur. **2021**(5), 13–16 (2021)
18. Minna, F., Massacci, F.: SoK: run-time security for cloud microservices. are we there yet?. Comput. Secur. 103119 (2023)
19. National Institute of Standards and Technology (NIST): National Vulnerability Database. https://nvd.nist.gov/vuln-metrics/cvss. Accessed 21 Mar 2024
20. Shopify: kubeaudit. GitHub (2023). https://github.com/Shopify/kubeaudit
21. Simonetto, S., Bosch, P.: Are we reasoning about cloud application vulnerabilities in the right way? In: 8th IEEE European Symposium on Security and Privacy (2023)
22. The MITRE Corporation: Common attack pattern enumeration and classification. https://capec.mitre.org/. Accessed 21 Mar 2024
23. The MITRE Corporation: Common Weakness Enumeration (CWE). https://cwe.mitre.org/. Accessed 21 Mar 2024
24. The MITRE Corporation: CVE. https://cve.mitre.org/. Accessed 21 Mar 2024

Improving Understanding of Misinformation Campaigns with a Two-Stage Methodology Using Semantic Analysis of Fake News

Sidbewendin Angelique Yameogo(✉)

Université de Bretagne Sud, Vannes, France
sidbewendin.yameogo@univ-ubs.fr

Abstract. Internet and social media are fueling the spread of disinformation on an unprecedented scale. Numerous tactics and techniques, such as Fake News, are employed to seek geopolitical advantages or financial gains. Many studies have focused on the automatic detection of Fake News, particularly using machine learning techniques. However, an informational attack often involves various vectors, targets, authors, and content. Detecting such an attack requires a global analysis of multiple Fake News instances. This research proposal aims to assist specialists, such as intelligence analysts or journalists responsible for combating disinformation, in better characterizing and detecting informational attacks.

We propose a framework based on a two-stage approach. The first stage involves extracting valuable knowledge from each Fake News using both Artificial Intelligence and Natural Language Processing (NLP) techniques. The second stage entails aggregating the collected information using data analysis methods to facilitate the characterization and identification of disinformation campaigns.

Keywords: Disinformation · Fake-news · Analysis · Conceptual Modeling

1 Introduction

Disinformation, amplified by social media, has emerged as a powerful weapon in contemporary information warfare [14]. In this environment conducive to information manipulation, a major player has arisen: Fake News. Evolving beyond mere vectors of misinformation, they have become a global point of interest, fueling crucial debates about their impact, rapid online spread, and challenges posed to an increasingly connected society [17].

The traditional approaches to combating Fake News often rely on binary classification techniques to distinguish between true and false information [9,10,17]. These techniques provide a simplistic characterization by considering each piece of information individually. Moreover, they offer little information about the specific reasons or characteristics that led to the classification of information

J. Araújo et al. (Eds.): RCIS 2024, LNBIP 514, pp. 121–130, 2024.
https://doi.org/10.1007/978-3-031-59468-7_14

as false. Thus, while these methods may determine whether information is true or false, they provide little insight into the intrinsic nature of falsehood. This poses a challenge when aggregating information to understand a disinformation campaign. Indeed, to effectively analyze a disinformation campaign, it is essential to deeply understand what characterizes information as false.

In this context, our approach stands out by proposing a thorough analysis, examining each piece of Fake News both individually and aggregatively. We are committed to exploring how semantically rich characterization of Fake News can play a crucial role in improving the identification and analysis of information campaigns. Our primary objective is to enhance the semantic characterization of Fake News and the detection of informational attacks, providing intelligence analysts and journalists involved in the fight against disinformation with a deeper understanding of Fake News.

To materialize this vision, we propose the development of an automated framework to establish a repository of Fake News. This repository will serve as an operational hub, leveraging tools from data science and artificial intelligence. These advanced instruments will enable sophisticated processes such as interrogation, visualization, and in-depth analysis of Fake News, thereby contributing to a better understanding and detection of informational attacks.

The paper is organized as follows. Section 2 presents the state of the art in the field of disinformation campaigns and Fake News, with a particular focus on definitions and detection methods, as well as the means used to analyze and respond to such informational attacks. Section 3 outlines the research objectives and questions. Section 4 presents the proposal and the approach we put in place to meet the objectives.

2 Related Works

Information warfare encompasses various concepts. In this section, we present some of them. Firstly, we examine the concept of disinformation campaigns. Then, we focus on the concept of Fake News, delving specifically into efforts related to the detection and characterization of Fake News. Finally, we explore methods of analyzing and, where possible, responding to an informational attack.

2.1 Disinformation Campaign

The rise of information warfare has bestowed a strategic dimension upon the dissemination of Fake News. NATO[1] describes information warfare as "an operation conducted to obtain an informational advantage over the adversary." It focuses on information, its manipulation, its flow, how it is protected or stolen, and how it is used [8]. In this context, a disinformation campaign is an integral part of information warfare. Fake News emerge as potent instruments in this dynamic, serving as projectiles of disinformation aimed at specific objectives. This challenge has intensified, especially during the 2016 US presidential

[1] NATO: North Atlantic Treaty Organization https://www.nato.int/.

elections, characterized by a widespread dissemination of false information [1]. The proliferation of Fake News in recent years has underscored the imperative to comprehensively analyze these phenomena. For instance, disinformation campaigns will often post overwhelming amounts of content with the same or similar messaging [4]. In this era where mastering and understanding disinformation campaigns have become crucial, examining these campaigns carefully lays the foundation for a stronger defense against disinformation.

2.2 Fake News Detection

In response to the increasing prevalence of disinformation campaigns, several initiatives have emerged with the objective of automating the detection of Fake News. Artificial intelligence has been extensively utilized, exploring approaches based on fact consistency, writing style, online propagation, and source credibility [7]. These research endeavors have heavily invested in the pursuit of practical and automated methods for the binary detection of Fake News online, with the aim of assisting users in discerning the truthfulness of information [9,10,17].

Despite the remarkable effectiveness of these models in identifying deceptive information, growing concerns have arisen regarding their transparency and explanatory capability [5,6,11]. Indeed, online users are often left in the dark about the precise functioning of these detection mechanisms.

Detecting online disinformation is a complex task that goes beyond the capabilities of a simple binary classification model. To develop a reliable detection algorithm, it is imperative to surpass the limitations of traditional binary models and identify specific distinctive features for machine learning [16]. These characteristics will enable us to understand Fake News more precisely and in context.

2.3 Fake News Characterization

Addressing the challenge of disinformation goes beyond merely detecting Fake News. A more nuanced characterization of the nature of Fake News is crucial. Recent research has delved into characterizing Fake News, identifying specific traits related to creators, targeted victims, content, and social context [2,16]. While the proposed characterization is relevant, it lacks reliance on an explicitly defined conceptual model. A conceptual model allows for defining a domain with specific and precise semantics, utilized by human users to facilitate communication, discussion, negotiation, and so on. Using a conceptual model enables a more informative definition of the Fake News domain and facilitate the construction of well-justified and explainable models for the detection and understanding of informational campaigns, which have been rarely available to date.

Research efforts such as [3] have proposed a departure from conventional approaches by providing a detailed characterization of what constitutes Fake News. They suggest an innovative approach using a conceptual model to characterize the content of Fake News. Drawing inspiration from the Explainable Artificial Intelligence (XAI) process [12], they guide the construction of a justified and explanatory model of Fake News generation.

However, the primary use of the conceptual model focuses on detecting Fake News and their automated generation, which differs from our approach. In our work, the conceptual model is a basis for characterizing fake news that we attempt to fill in order to enable more meaningful aggregation. This is a first step towards identifying characteristic traits of fake news that will facilitate aggregate processing to detect disinformation campaigns.

2.4 Disinformation Analysis and Response Measures

Beyond the detection and characterization of Fake News, combating information warfare requires proactive and adaptive measures. Recognizing that risk reduction targeting populations requires heightened collaboration between cyber disciplines and other specialties, the DISARM Foundation[2] has developed a matrix dedicated to private/public entities collaborating to counter disinformation [14]. This matrix, operating within an open-source and secure framework, outlines an information manipulation campaign through tactics, techniques, and procedures, similar to the Mitre ATT&CK matrix [13] for cyber attacks.

While DISARM offers advantages, including its adaptability to different incident scales, usable threat matrices, and continuous use since 2019, it does not rely solely on automation. This characteristic underscores the importance of human intervention to maintain system relevance and responsiveness. In this article, our approach aligns with this perspective by focusing on the aspects of post-automated analysis of "Fake News" classification. The use of structured data resulting from this analysis will enable the analyst to establish a connection with the DISARM matrix, providing tangible means to identify an ongoing form of attack. This contributes to a better understanding of long-term disinformation campaigns.

3 Research Objectives

This section is devoted to presenting the key objectives and rationale of our research, along with the specific research questions that motivate it.

3.1 Research Objectives

Related works highlight the prevalence of research focused on the binary detection of Fake News. Despite their effectiveness in identifying misleading information, these models have raised growing concerns about transparency and explainability. The results of these approaches provide a set of labeled news (fake or not). We propose to extend these results by semantically enriching the concept of Fake News. Thus, we go beyond binary classification to provide a comprehensive and contextual characterization of Fake News. By applying techniques of

[2] DISARM: Disinformation Analysis and Response Measures, https://disarmframework.herokuapp.com/.

artificial intelligence and data science to this knowledge base of fake news, we aim to provide a comprehensive analysis tool contributing to the understanding of disinformation campaigns.

Structuring the information contained in Fake News within a more refined semantic representation makes it possible to aggregate them in order to identify links between news. This capability helps users in identifying coordinated informational attacks and establishing connections with dedicated frameworks such as the DISARM matrix. In doing so, our target audience includes intelligence analysts and journalists, assisting them in their investigative work and the identification of disinformation campaigns.

3.2 Research Questions

In this context, two research questions emerge.

RQ1 How should we improve understanding of Fake News by characterizing them with semantically rich elements?
This question aims to address the inherent challenge of limited characterization of Fake News. By exploring semantic approaches, we seek to go beyond binary classification of news (fake or not) to obtain a more precise representation of false information.

RQ2 How can this characterization contribute to a better identification and analysis of an disinformation campaign?
We look at the possibility of understanding and analyzing an informational attack carried out as part of a disinformation campaign using multiple Fake News.

By addressing these questions, we aspire to develop an innovative method that strengthen the characterization of Fake News and illuminate the understanding of disinformation campaigns, contributing to more effective measures against these complex informational threats.

4 Contributions and Advances

In this section, we introduce the main principles guiding our approach by firstly defining a two-stage methodology that we would integrate into an analysis framework. Next, we describe our preliminary results in developing such a framework. Finally, we outline the next steps we will undertake.

4.1 A Two-Stage Methodology for Analyzing Fake News

As mentioned in Sect. 2, the majority of existing works on Fake News have focused on detecting it. Our work does not aim to increase Fake News detection capability. We operate under the hypothesis that existing Fake News detection approaches are capable of indicating whether a news article is fake or not. Our plan is to use such results as a starting point, beginning with identified Fake News, and applying additional treatments to improve its characterization. With this objective in mind, the work of [3] proposed a common understanding of what constitutes Fake News by introducing a precise conceptual model. The problem to be addressed is how to extract data from a news article identified as false, in order to conduct a semantic analysis. Ideally, we want to use this conceptual model. However, extracting data (automatically or not) from a news article conforming to this conceptual model is a challenging task. Moreover, once data is extracted following the conceptual model, processing it from the perspective of an informational attack is difficult due to both the high number and diversity of Fake News to be considered.

To address such a need and respond to the research questions outlined in Sect. 3.2, we have adopted a two-phase processing approach, as illustrated in Fig. 1. The first phase, named "Unitary Level", focuses on the individual and independent processing of each Fake News to address the first research question. The goal is to enhance our understanding of Fake News by characterization it with semantically rich elements. This involves expanding knowledge about each case of Fake News unitary, thereby enriching existing databases with features that we consider relevant based on the conceptual model.

In this process, the use of artificial intelligence (AI) and NLP tools is considered. These tools have been selected for their ability to efficiently process the complex, unstructured data that is characteristic of our field of study, as well as for their potential to provide meaningful, actionable insights. For example, we rely on named entity extraction, which involves the use of algorithms based on recurrent neural networks (RNNs) or transformer models such as BERT. This technique will enable us to identify and extract important entities such as people, places and organizations mentioned in Fake News articles. We plan to use probabilistic topic modeling algorithms such as Latent Dirichlet Allocation (LDA) or Non-Negative Matrix Factorization (NMF). These approaches will enable us to discover the underlying themes and recurring topics in Fake News. We will also rely on sentiment analysis techniques, such as Convolutional Neural Networks (CNN), Support Vector Machines (SVM), ... to understand the emotions and reactions associated with Fake News.

The conceptual model assumes a central role in guiding the extraction process towards the fullest and richest features, facilitating a thorough characterization of the Fake's information.

Furthermore, most Fake News is part of organized attacks that can be carried out over the long term. Thus, the independent study of each Fake News item does not provide a complete and exhaustive overview of the attack, highlighting the limitations and inadequacy of this isolated approach. To address this complexity, the second phase, named "Aggregation Level" in Fig. 1, involves the aggregated processing of Fake News. In this context, human intervention is of crucial importance to enhance the data analysis process by enabling feedback and updating of the conceptual model. Individuals contribute in real-time by providing contextual information and enriching the database, thereby improving the quality and diversity of data available for analysis. This continuous interaction between human experts and the instantiated conceptual model plays an essential role in the adaptation and evolution of the system to better understand and counter information attacks.

This step aims to address the second research question, focused on contributing to a better understanding, identification, and analysis of an information campaign. The objective of this phase is to utilize the results of individual analyses stored in the instantiated model (see Fig. 1) to gain a deeper understanding of information attacks.

This phase of Fake News analysis employs sophisticated data analysis techniques such as *Factorial Analysis of Mixed Data (FAMD)*, *clustering*, *ensemble queries*, *distance calculation*, and *spatial vision*. We will use Mixed Data Analysis Factorization (FAMD) to effectively manage datasets containing both numerical and categorical variables. This technique will enable us to uncover hidden patterns in the data by reducing its dimensionality while preserving the relationships between variables. Clustering will be employed to group similar instances together, allowing us to identify common characteristics among false news articles. Ensemble queries will assist us in extracting valuable insights from multiple sources simultaneously, thus enhancing the scope of our analysis.

Furthermore, distance calculation techniques will be utilized to measure the dissimilarity between data points, facilitating the identification of outliers and anomalies in the dataset. Spatial vision techniques will provide a visual representation of the data, enabling us to discern spatial patterns and relationships that may not be apparent through simple numerical analysis.

As far as Disarm is concerned, the incorporation of our tool offers beneficial potential for the analyst when identifying attacks. As a reminder, Disarm is a matrix dedicated to analyzing and defending against informational attacks. To illustrate this with a concrete example, a possible tactic for launching an informational attack would be to apply tactic T009 "*Create fake experts*", followed by tactic T0045 "*Use fake experts*". The detection of a non-existent or non-competent authority (fake experts) by our tools could thus signal the use of this attack tactic, enabling the analyst to suggest specific countermeasures. In this context, the visualization of Fake News narratives generated by these complex analyses adds an extra layer of depth to the understanding of information warfare. It enables the exploration of concrete questions such as the spatial distribution of Fake News, the emergence of specific clusters, dynamic evolution based on critical variables, and detecting trends in spreading misinformation. Moreover, integrating these analyses could provide analysts with a more comprehensive and contextual view of disinformation patterns, thereby strengthening the DISARM frameworks. This would enable analysts to better comprehend the behaviors of incident creators, understand how specific disinformation tactics propagate, and formulate tailored responses to disinformation attacks.

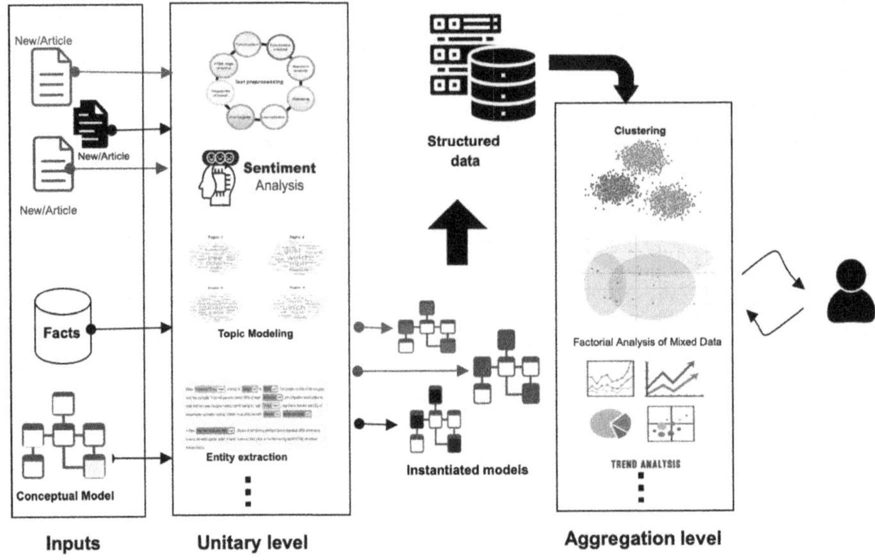

Fig. 1. Illustration of the proposed two-phase approach

4.2 First Results

Currently, we are focusing on the first processing level of our approach, namely the Unitary level (see Fig. 1). To conduct it, we chose at first to use a database called the "LIAR dataset" [15]. This database is commonly employed in the field of Fake News detection. It was created based on statements sourced from the fact-checking website *Politifact*. These statements are annotated with six truthfulness categories, ranging from "True" to "Pants on Fire", including "Mostly True", "Half True", "Mostly False", and "False". However, for our specific objective, we only utilized data labeled as false. After selecting the data, we undertook an initial pre-processing step to ensure the consistency and quality of the information. This process involved using Python programming tools, including libraries such as scikit-learn, NumPy, and Pandas.

Simultaneously, within the framework of this experimentation, we utilized the conceptual model defined in [3] to guide the process. This model holds a central position in our approach, providing the necessary structure for feature extraction. Automated feature extraction techniques have been employed, with a focus on NLP tools to delve into the semantic and linguistic aspects of the data, thereby offering an enriched perspective on Fake News. By orchestrating these various chunk of treatments cohesively, our initial experimentation laid the foundation for a nuanced understanding of Fake News.

Pre-labeled data, coupled with advanced feature extraction techniques and guided by our conceptual model, enabled the capture of the semantic complexity of misleading information. We believe that these results form a robust platform for the subsequent phases of our project, aiming to integrate this knowledge into a broader conceptual model.

For example, the database we utilized encompasses three key features: the title of the Fake News, the content of the false article, and the author who created this misinformation. When we input a piece of Fake News into our initial unitary processing phase, relying on the conceptual model and utilizing the aforementioned text analysis tools, we generate an output instance of the Conceptual Model filled with additional characteristics, such as the countries mentioned in the article (facilitating the localization of targeted countries), the authorities cited in the text to lend credibility to the false news, the mentioned dates to position the false news in time, and so on. Subsequently, this instance is stored in a database. This database will serve as a stating point for the next step, namely the aggregation step.

4.3 Discussion and Next Steps

Our approach to enhancing the analysis of informational attacks has been guided by the use of a conceptual model. This model holds a pivotal position at the core of our approach, providing an essential structured framework for extracting crucial features necessary for understanding and characterizing Fake News.

In our initial experiments, we found that NLP techniques could be used to extract information from Fake News that corresponded to certain semantic elements in the model. We are continuing in this direction in order to identify new additive techniques aimed at identifying the remaining semantic elements. However, it is possible that the current state of the art will not allow us to identify all the elements expressed in the model. In this context, it may be necessary to move towards a semi-automated approach in which a human operator can intervene to compensate for the shortcomings of NLP approaches. We also would only use a sub-model of the original conceptual model.

As for the aggregation part of our approach, we propose to build a repository containing the fake news processed in the first phase. This will be structured according to the conceptual model obtained. It will then be possible to apply set-type queries to it, for example. We are actually building such a repository. Later, we'll identify higher-level, high value-added queries that can be applied to the fake news repository. We are thinking in particular of data analysis techniques such as *Factorial Analysis of Mixed Data (FAMD)* or *clustering.*

Finally, it will be necessary to deeper study how creating correspondences between DISARM and the result of the aggregative part of our work. Indeed, this allow to better guide the analyst to the identification of the different steps of an informational attack, following the DISARM guidelines.

Supervision

This work is realized under the supervision of Regis Fleurquin, associate professor at Université de Bretagne Sud, Wassila Ouerdane full professor at Ecole

Centrale-Supelec and Nicolas Belloir, associate professor at Academy Militaire de St Cyr Coëtquidan. It has never been presented at a doctoral symposium.

References

1. Albright, J.: The #election2016 micro-propaganda machine. Medium (2016)
2. Ansar, W., Goswami, S.: Combating the menace: a survey on characterization and detection of fake news from a data science perspective. Int. J. Inf. Manag. Data Insights **1**(2), 100052 (2021)
3. Belloir, N., Ouerdane, W., Pastor, O.: Characterizing fake news: a conceptual modeling-based approach. In: Ralyté, J., Chakravarthy, S., Mohania, M., Jeusfeld, M.A., Karlapalem, K. (eds.) ER 22. LNCS, vol. 13607, pp. 115–129. Springer, Cham (2022). https://doi.org/10.1007/978-3-031-17995-2_9
4. Cybersecurity and Infrastructure Security Agency (CISA): Tactics of disinformation (2022). https://www.cisa.gov/sites/default/files/publications/tactics-of-disinformation_508.pdf
5. Gadek, G., Guélorget, P.: An interpretable model to measure fakeness and emotion in news. Procedia Comput. Sci. **176**, 78–87 (2020)
6. Maass, W., Castellanos, A., Tremblay, M.C., Lukyanenko, R., Storey, V.C.: AI explainability: embedding conceptual models. In: Proceedings of the 43rd International Conference on Information Systems, ICIS 2022 (2022)
7. Molina, M.D., Sundar, S.S., Le, T., Lee, D.: "fake news" is not simply false information: a concept explication and taxonomy of online content. Am. Behav. Sci. **65**(2), 180–212 (2021)
8. NATO: Media - (dis)information - security (2005). https://www.nato.int/nato_static_fl2014/assets/pdf/2020/5/pdf/2005-deepportal4-information-warfare.pdf
9. Phan, H.T., Nguyen, N.T., Hwang, D.: Fake news detection: a survey of graph neural network methods. Appl. Soft Comput., 110235 (2023)
10. Rohera, D., et al.: A taxonomy of fake news classification techniques: survey and implementation aspects. IEEE Access **10**, 30367–30394 (2022)
11. Shu, K., Cui, L., Wang, S., Lee, D., Liu, H.: dEFEND: explainable fake news detection. In: Proceedings of the 25th ACM SIGKDD International Conference on Knowledge Discovery & Data Mining, pp. 395–405 (2019)
12. Spreeuwenberg, S.: AIX: Artificial Intelligence Needs Explanation (2019)
13. Strom, B.E., Applebaum, A., Miller, D.P., Nickels, K.C., Pennington, A.G., Thomas, C.B.: MITRE ATT&CK: design and philosophy. Technical report, MITRE Corporation (2018)
14. Terp, S., Breuer, P.: DISARM: a framework for analysis of disinformation campaigns. In: 2022 IEEE Conference on Cognitive and Computational Aspects of Situation Management (CogSIMA), pp. 1–8. IEEE (2022)
15. Wang, W.Y.: "Liar, Liar pants on fire": a new benchmark dataset for fake news detection. In: Proceedings of the 55th Annual Meeting of the Association for Computational Linguistics, pp. 422–426. Vancouver, Canada, July 2017
16. Zhang, X., Ghorbani, A.A.: An overview of online fake news: characterization, detection, and discussion. Inf. Process. Manag. **57**(2) (2020)
17. Zhou, X., Zafarani, R.: A survey of fake news: fundamental theories, detection methods, and opportunities. ACM Comput. Surv. (CSUR) **53**(5), 1–40 (2020)

Automated Scriptless GUI Testing Aligned with Requirements and User Stories

Mohammadparsa Karimi(✉)

Open Universiteit, Heerlen, The Netherlands
parsa.karimi@ou.nl

Abstract. Testing is an essential phase of software development to evaluate the quality of the product. Scriptless testing is a prominent technique that makes this phase efficient. However, there is a research gap in automating the testing process from the requirements. In this research we want to propose an innovative approach: Automated Scriptless GUI Testing Aligned with Requirements and User Stories. Using the open-source GUI testing tool, TESTAR, we want to propose an AI-powered tool that enables TESTAR to test software against specified requirements and user stories.

Keywords: GUI Testing · Scriptless Testing · Requirements Testing

1 Introduction

Testing is the most common technique used to assess the quality of software products [12]. In any software development process, requirements serve as the foundations that guide developers and testers in understanding what the software should do and how it should perform. To ensure that the software meets the customer's needs and remains safe from potential threats, it's important to have a clear understanding of the system's requirements [3]. Nevertheless, the most serious software failures are often caused by the lack of understanding and testing the requirements, not coding errors [22].

Research in requirements engineering has been done with several perspectives, especially in terms of agile development [5], human collaboration [4] and its relationships with innovation theories [13]. Despite the acknowledged impact of agile development on requirements engineering and testing [14], there is a significant lack of studies describing how requirements are specified, managed and tested [8,21].

Despite ongoing efforts to bridge the gap in requirements testing, many challenges remain. For example, model-based testing, which has received considerable attention from researchers, seeks to improve testing methods for evaluating software against its requirements [17,20]. However, testing remains costly due to maintenance challenges and the complex task of developing test models [10]. Another example of efforts in this area is automated acceptance testing using

J. Araújo et al. (Eds.): RCIS 2024, LNBIP 514, pp. 131–140, 2024.
https://doi.org/10.1007/978-3-031-59468-7_15

Requirements Specification Language (RSL) language from requirements [15]. In [15], the solution is to adopt a model-based approach that aligns requirements with tests, including both test cases and low-level automated test scripts. While these measures have succeeded in speeding up the testing process, it's important to recognise that human involvement remains essential, because even if the test model is created automatically, there should be a human to perform it. In agile development, where speed and efficiency are paramount [23,27], there is a growing need for even faster and more efficient methods of evaluating software against its requirements and user stories.

In the field of software testing, there's a technique known as automated scriptless testing. Scriptless approaches work by generating sequences of user actions on-the-fly at runtime to explore the system under test (SUT). This involves dynamically selecting and executing available actions within the detected graphical user interface (GUI) states, allowing adaptive and context-aware exploration of the SUT's functionality without relying on pre-defined scripts [16].

In this research, our aim is to: ***Extend scriptless testing methods with AI-enabled components that can learn the best way to select actions to automatically test requirements and user stories.*** We aim to use the scriptless appraoach TESTAR [24] since it has a modular architecture and scalable flow to add our extension to enter the functional requirements testing paradigm.

This paper is structured as follows. Section 2 describes the background, Sect. 3 presents our solution, research goals and domain knowledge so far. Finally, in Sect. 4, the plans for validation and evaluation are discussed.

2 Background

Scriptless testing is based on a paradigm that shifts the challenge from the creation and maintenance of scripts to the creation of intelligent agents that need to determine the most effective ways for testing the software [9].

A prominent example of this approach is the TESTAR tool [24]. TESTAR tests software using a simple process with four main steps. First, it starts up the software in its initial state and detects the state. Then it derives all possible actions and selects one. After executing the action, the software changes to a new state. Finally, TESTAR oracles check that the new state works well. This tool is built in a modular way to facilitate the extensions [24]. The TESTAR high level plugin architecture is shown in Fig. 1. TESTAR selects actions using an action selection mechanism (ASM), such as random or Reinforcement Learning (RL) algorithms, allowing it to explore software and uncover software defects such as crashes and freeze. In TESTAR, ASMs receive a list of possible actions and select one from the list. If the mechanism is a random selection process, the action is chosen randomly. Alternatively, if the mechanism contains an RL agent, it will select the action with the goal of exploring more states in the software. The operation of the different ASMs in TESTAR is illustrated in Fig. 2.

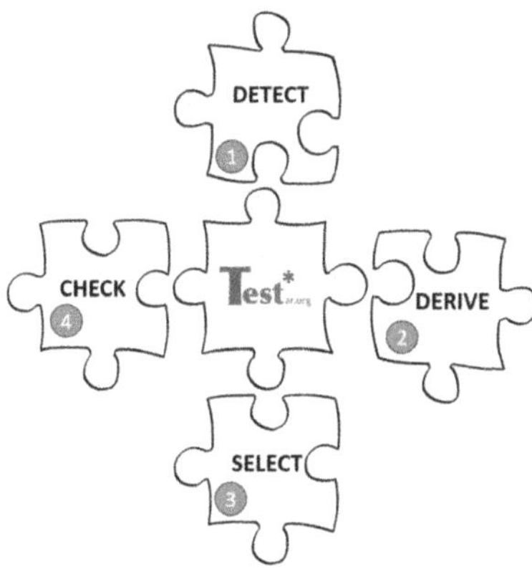

Fig. 1. TESTAR flow and high-level architectural modules

TESTAR has demonstrated its effectiveness in various industrial case studies. The use of TESTAR in real-world scenarios has validated its performance and its ability to detect and address potential problems [1,2,18].

Nevertheless, with the current ASMs implementation, it is impossible to know whether TESTAR has tested certain functional requirements or user stories, because requirements testing involves verifying that the software meets specified criteria and fulfills its intended purpose. To do this, a more targeted and informed approach for testing is needed, where the tool has knowledge of the underlying requirements and edge cases, and assesses whether the software meets those expectations. For example, in the case of a banking application as a SUT, TESTAR can currently check various parts of the software for crashes or freezes by executing actions. However, if the goal is to test the banking application against a specific requirement, such as money transfer, TESTAR needs to have knowledge of the SUT domain. This knowledge is crucial to detect how to navigate the software from its initial state to the money transfer page. In addition, TESTAR needs the knowledge to verify the test results and determine the success or failure of the funds transfer process.

3 Research

We advocate TESTAR can be used as baseline for implementation of automated requirements testing. By adding an intelligent action selection mechanism and smart test oracles, it is possible to evolve TESTAR into an automated scriptless testing tool for evaluating software against the requirements.

To accomplish the automated functional requirements testing, we need to implement specific improvements in the scriptless testing tool TESTAR, which lead to the research objectives of this thesis:

- **Objective 1.** Create behavioral software models: To consider the functional requirements expressed in user stories, TESTAR needs knowledge about the

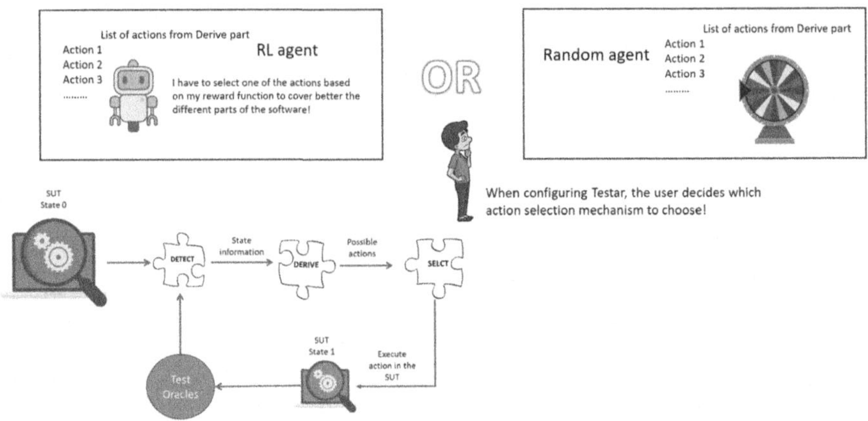

Fig. 2. Representation of the selection mechanisms in TESTAR

software domain. This knowledge can be acquired by creating behavioral models derived from the software during manual testing or the execution of automated script runs.
- **Objective 2.** Develop intelligent action selectors using AI: Intelligent action selectors powered by artificial intelligence need to be developed. These selectors should guide the testing tool to achieve the desired state of the user story by executing a sequence of actions.
- **Objective 3.** Create intelligent test oracles: The creation of intelligent testing oracles is necessary to check and report the behavior of the software during testing against specific user stories, ensuring compliance with the acceptance criteria. In this thesis we want to use large language models as smart test oracles.
- **Objective 4.** Defining test adequacy criteria: It is essential to define precise test adequacy criteria in order to effectively evaluate our tool. These criteria will serve as benchmarks to ensure accuracy and relevance in achieving our testing objectives.

3.1 Create Behavioral Software Models

To comprehend the functional requirements articulated in user stories, TESTAR requires familiarity with the software domain.

To do that, we propose the development of a new mode for TESTAR, which we call the "listening mode". This mode will be designed to observe both human interactions and the execution of automated test scripts, in order to generate state models based on these executions. The workflow of this concept is illustrated in Fig. 3.

This way TESTAR will be capable of creating a model for the specific part of the SUT the tester is examining, including the interactions that testers perform

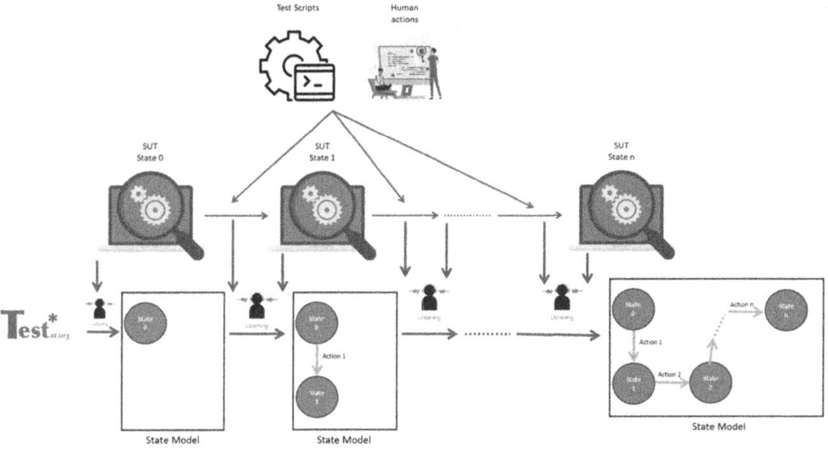

Fig. 3. Listening mode operation flow

[11]. This model will keep a record of all the different states explored during a test run. The resulting state model will help the AI action selector choose which actions to take.

3.2 Develop Intelligent Action Selectors Using AI

We will develop an AI-based intelligent action selector capable of follow a user story and make decisions to achieve the goal outlined in that story. The proposed action selection mechanism should be based on known testing methodologies and domain knowledge of the SUT. Our approach involves designing and testing various AI algorithms to create an effective action selector capable of faithfully reproducing the goals defined in a user story.

Large Language Models. First, we will use a pre-existing language model as an action selector to evaluate its effectiveness in selecting actions. In this particular scenario, our goal is to use the language model as an online agent capable of selecting actions based on the provided user story and information about each state of the SUT. Two inputs for the language model in this section are the user story and the potential actions from the current state of the SUT. Our goal in this phase is to:

- Evaluate different language models to compare their ability to successfully test user stories.
- Evaluate different prompts to determine which are more effective in selecting actions for functional requirement testing.

Figure 4 illustrates the flowchart detailing the concept of using an existing language model as an action selector for testing software against user stories.

The process starts with the system under test in its initial state. This state is provided to the recognition module of TESTAR, which generates a list of possible actions. The list, together with the state information, is then passed to the language model along with the user story. The language model then makes a decision about the next action. If the model determines that the software has reached the target state defined in the user story, it passes this information, along with acceptance criteria, to the test oracle. If the model concludes that the software has not reached the target state, it selects the best action and returns it to TESTAR's selection module. TESTAR, in turn, moves the software to the next state by executing the selected action, and this iterative process continues, with information feedback to the detection module for subsequent iterations.

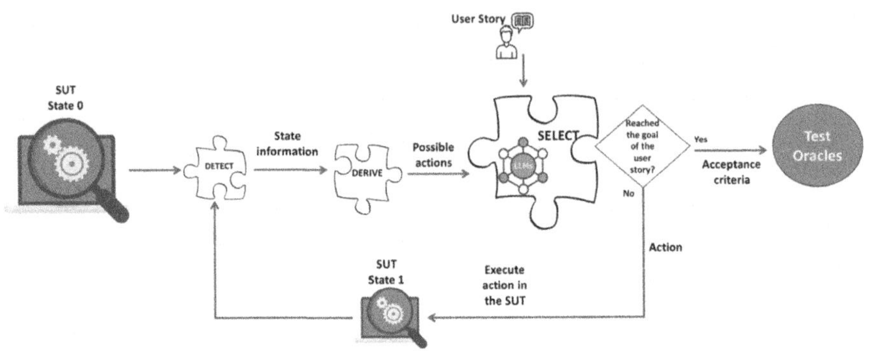

Fig. 4. Using an existing language model flow chart

Language models are commonly trained across various domains and may not possess specific expertise in the field of software testing. Our next step is to refine and fine tune the model, creating a language model that not only comprehends the complexities of software testing but also effectively addresses the challenges presented by various test scenarios.

To implement this method effectively, We need a data set for fine-tuning the model. The generation of this dataset involves the creation of state models from test artifacts. Considering that in **Objective 1** "Create behavioral software models" we aim to integrate the Listening Mode in TESTAR, we can at the same time capture the execution of the different types of tests, including both automated test script executions and manual tests rooted in user stories and system requirements to create the dataset. Figure 5 illustrates the process of creating the dataset and fine-tuning the model.

Deep Neural Network Agent. We plan to use a pre-trained deep learning agent. This agent gets its training data from test examples made with TESTAR Listening Mode, as we mention above.

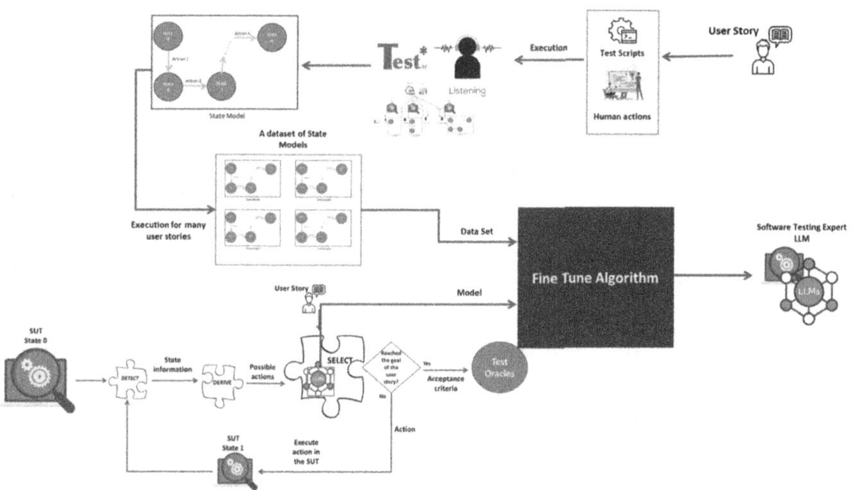

Fig. 5. Fine tune LLM with state models

The agent comes to the task already knowing things from its initial training on a set of test examples. Therefore, any decisions or problem solving the agent does during the testing task is based only on what it has learned during the training process.

By using this deep learning agent, we aim to assess its performance in comparison with a language model approach. To do that, we aim to build a dataset using industrial test artifacts. After that, we plan to train an offline deep learning agent on this dataset. Based on the knowledge gained during training, the agent can make informed decisions about the next action to determine the achievement of the user story state.

3.3 Create Intelligent Test Oracles

Test oracles are like inspectors in the validation process. Oracles check whether the conditions specified in a user story's acceptance criteria are met. The oracles use a systematic method, following clear steps to intelligently figure out how to assess the functional conditions of a system. Oracles can be adapted to different situations and ensure a thorough evaluation about the software meeting the user expectations.

For example, when testing a banking application against the requirement to transfer funds, the first action selection is to perform a transfer. The oracle then needs to perform further actions to navigate to the transaction page. The oracle then checks the state of the transaction to determine its success or failure. To do this, the oracle requires knowledge of the SUT, the associated user story and the acceptance criteria. This knowledge is essential for the oracle to make informed decisions about whether the software meets the specified requirements.

3.4 Defining Test Adequacy Criteria

A critical factor in assessing how well the SUT will perform is knowing how good a test is. While a well-designed test provides confidence in the correctness of the product, it's important to remember what computer science pioneer Edgar Dijkstra said back in the 1970s - complete correctness cannot be proven by testing alone [19]. In software testing, practitioners often use alternative measures such as code coverage or mutation rate to check whether a test is good enough.

Although there are many standards in the field of software testing, implementing a metric and measuring that in practice remains challenging. In our approach, we want to introduce novel ways of measuring how well a test covers things, closely related to the behavioral models we made in **Objective 1**, following the methodological approach to define measures presented in [6,7]. These novel measures will give a detailed view of how well a test run meets the various acceptance criteria associated with user stories. For example, we plan to set criteria based on the coverage of various language components identified in the traceability matrix, which match well with the language used in the user stories.

Moreover, we aim to create criteria that operate at a higher level than traditional testing metrics and provide a more complete assessment of test quality. To ensure that the proposed criteria are effective, we plan to conduct empirical studies. These studies will not only test whether the proposed criteria are useful, but will also provide real evidence of how well they assess the quality of the testing process. This comprehensive approach should improve our understanding of test quality and provide valuable insights into the wider field of software testing.

4 Plan for Validation and Evaluation

The research strategy behind follows the principles of design science research [26]. Design science proposes a cycle of four main steps: problem investigation, treatment design, treatment validation, and treatment implementation. We plan to use the design cycle with the four engineering steps during the whole thesis. After identifying the problem, we research and design different solutions. After validation of the design, we will implement the tools. Moreover, we will evaluate the tools' performance using specific metrics, and refine our approach accordingly.

To evaluate work, we need to design experiments for each task to determine its usefulness and effectiveness. For the experiments, we need a detailed plan which considers:

- Identify the SUT to be tested and its platform (such as web, Android, etc.). The SUT can be from industry and open source platforms.
- Define the evaluation variables considering the measures defined in **Objective 4**.
- Design and implement the architecture needed for the experiments.
- Identify the proper analysis techniques to understand the results obtained.
- Comparing our results with other approaches.

We are currently in the initial phase of the project, and this proposal was written and submitted during the first year of the PhD program. During the following years of the project we will concentrate on the evaluation of the tools designed and the development of the tools. We will use TESTAR and the TESTAR infrastructure to carry out these experiments and for evaluating the results, we will use a methodological framework [25] which was introduced for evaluating software testing techniques.

In line with open science principles, we intend to develop the extensions of tool in the TESTAR GitHub repository (https://github.com/TESTARtool) and publish open-access publications.

Acknowledgement. This work has been funded by the NWO OTP project AUTOLINK (19521) and under co-supervision of Beatriz Marín from Universitat Politècnica de Valencia.

References

1. Bons, A., Marín, B., Aho, P., Vos, T.E.: Scripted and scriptless GUI testing for web applications: an industrial case. Inf. Softw. Technol. **158**, 107172 (2023)
2. van der Brugge, A., Pastor-Ricós, F., Aho, P., Marín, B., Vos, T.E.: Evaluating testar's effectiveness through code coverage. Actas de las XXV Jornadas de Ingeniería del Software y Bases de Datos (JISBD) 1–14 (2021)
3. Burkin, V.: Mitigating risks in software development through effective requirements engineering. arXiv preprint arXiv:2305.05800 (2023)
4. Conboy, K., Coyle, S., Wang, X., Pikkarainen, M.: People over process: key challenges in agile development. IEEE Softw. **28**(4), 48–57 (2010)
5. Dingsøyr, T., Nerur, S., Balijepally, V., Moe, N.B.: A decade of agile methodologies: towards explaining agile software development (2012)
6. Giachetti, G., Marín, B., Franch, X.: Using measures for verifying and improving requirement models in MDD processes. In: 14th International Conference on Quality Software, pp. 164–173. IEEE (2014)
7. Giachetti, G., Marín, B., López, L., Franch, X., Pastor, O.: Verifying goal-oriented specifications used in model-driven development processes. Inf. Syst. **64**, 41–62 (2017)
8. Inayat, I., Salim, S.S., Marczak, S., Daneva, M., Shamshirband, S.: A systematic literature review on agile requirements engineering practices and challenges. Comput. Hum. Behav. **51**, 915–929 (2015)
9. Jansen, T., et al.: Scriptless GUI testing on mobile applications. In: 22nd International Conference on Software Quality, Reliability and Security (QRS), pp. 1103–1112. IEEE (2022)
10. Marín, B., Gallardo, C., Quiroga, D., Giachetti, G., Serral, E.: Testing of model-driven development applications. Softw. Qual. J. **25**, 407–435 (2017)
11. Marín, B., Giachetti, G., Pastor, O., Abran, A.: Interaction models matter in the evaluation of quality of conceptual models. In: 13th International Conference on Quality Software, pp. 382–389. IEEE (2013)
12. Myers, G.J., Badgett, T., Thomas, T.M., Sandler, C.: The art of software testing, vol. 2 (2004)
13. Nerur, S., Balijepally, V.: Theoretical reflections on agile development methodologies. Commun. ACM **50**(3), 79–83 (2007)

14. Paetsch, F., Eberlein, A., Maurer, F.: Requirements engineering and agile software development. In: 12th International Workshops on Enabling Technologies: Infrastructure for Collaboration Enterprises (WETICE), pp. 308–313. IEEE (2003)
15. Paiva, A.C., Maciel, D., da Silva, A.R.: From requirements to automated acceptance tests with the RSL language. In: Damiani, E., Spanoudakis, G., Maciaszek, L. (eds.) ENASE 2019. CCIS, vol. 1172, pp. 39–57. Springer, Cham (2020). https://doi.org/10.1007/978-3-030-40223-5_3
16. Pastor Ricós, F., Slomp, A., Marín, B., Aho, P., Vos, T.E.: Distributed state model inference for scriptless GUI testing. J. Syst. Softw. **200**, 111645 (2023)
17. Pérez, C., Marín, B.: Automatic generation of test cases from UML models. CLEI Electron. J. **21**(1), 3-1 (2018)
18. Prasetya, I., et al.: An agent-based approach to automated game testing: an experience report. In: 13th International Workshop on Automating Test Case Design, Selection and Evaluation, pp. 1–8 (2022)
19. Randell, B.: Software engineering in 1968, pp. 1–10 (1979)
20. dos Santos, J., Martins, L.E.G., de Santiago Júnior, V.A., Povoa, L.V., dos Santos, L.B.R.: Software requirements testing approaches: a systematic literature review. Requirements Eng. **25**, 317–337 (2020)
21. Schön, E.M., Thomaschewski, J., Escalona, M.J.: Agile requirements engineering: a systematic literature review. Comput. Standards Interfaces **49**, 79–91 (2017)
22. Somers, J.: The coming software apocalypse. Atlantic **26**, 1 (2017)
23. Vargas, N., Marín, B., Giachetti, G.: A list of risks and mitigation strategies in agile projects. In: 40th International Conference of the Chilean Computer Science Society (SCCC), pp. 1–8. IEEE (2021)
24. Vos, T.E., Aho, P., Pastor Ricos, F., Rodriguez-Valdes, O., Mulders, A.: testar-scriptless testing through graphical user interface. Softw. Test. Verif. Reliab. **31**(3), e1771 (2021)
25. Vos, T.E., Marín, B., Escalona, M.J., Marchetto, A.: A methodological framework for evaluating software testing techniques and tools. In: 2012 12th International Conference on Quality Software, pp. 230–239. IEEE (2012)
26. Wieringa, R.J.: Design Science Methodology for Information Systems and Software Engineering. Springer, Heidelberg (2014). https://doi.org/10.1007/978-3-662-43839-8
27. Yanjari, I., Marín, B., Giachetti, G.: An open-source framework for cross-platform testing in agile projects. In: 41st International Conference of the Chilean Computer Science Society (SCCC), pp. 1–8. IEEE (2022)

Towards a Cybersecurity Maturity Model Specific for the Healthcare Sector: Focus on Hospitals

Steve Ahouanmenou(✉)

Faculty of Economics and Business Administration, Department of Business Informatics and Operations Management, Ghent University, Ghent, Belgium
steve.ahouanmenou@ugent.be

Abstract. The intersection of healthcare and technology has brought unprecedented advancements, improving patient care, and enhancing operational efficiency. However, this integration has also exposed the healthcare sector to significant cybersecurity challenges. With the increasing digitization of patient records and the reliance on interconnected systems, healthcare organizations are becoming attractive targets for malicious actors seeking to exploit vulnerabilities for financial gain or to disrupt critical healthcare services. Our main contribution is a cybersecurity maturity level specific to the healthcare sector with a focus on hospital; based on rigorous Research Science Design Methodology. In other words, this research aims to investigate and address the multifaceted cybersecurity issues within the healthcare sector, focusing on hospitals, analyzing their cybersecurity profiles, proposing effective ways to accelerate cyber risks assessment in order to safeguard patient data, maintain system integrity, and ensure the continuity of healthcare services.

Keywords: Healthcare · Information Security · Risk Assessment · Maturity level

1 Problem Statement

The intersection of healthcare and technology has witnessed transformative changes in recent years, ushering in an era of digitization that promises improved patient outcomes and operational efficiencies. Electronic Health Records (EHRs), telemedicine platforms, medical devices, and data analytics tools have become integral components of modern healthcare delivery [1]. However, this digital transformation has exposed the healthcare sector to unprecedented information security and privacy challenges, threatening the confidentiality, integrity, and availability of sensitive patient information.

The escalating integration of digital technologies within healthcare infrastructures has created a vast attack surface for malicious actors, ranging from financially motivated hackers to state-sponsored cyber espionage groups [2]. The implications of successful cyber-attacks on healthcare organizations extend far beyond the compromise of patient data. Ransomware attacks, for instance, have the potential to disrupt critical medical

J. Araújo et al. (Eds.): RCIS 2024, LNBIP 514, pp. 141–148, 2024.
https://doi.org/10.1007/978-3-031-59468-7_16

services, leading to life-threatening consequences for patients. Moreover, the clandestine nature of cyber threats and the value of healthcare data on the black market make the sector an attractive target [3].

The COVID-19 pandemic further accelerated the adoption of telehealth and remote patient monitoring, amplifying the importance of secure digital communication channels and resilient healthcare infrastructures [4]. The rapid deployment of these technologies, however, often outpaced the implementation of robust information security and privacy measures, leaving vulnerabilities that threat actors exploit. Previous academic studies have been conducted on how to facilitate access and make use of patient data [5]. For instance, a review on cybersecurity in the healthcare sector aimed at improving US healthcare quality and reducing cost [6]. Another paper discussed the information security and privacy in hospitals focused on Electronic Health Records (EHRs) systems [7]. Other studies related to cybersecurity in hospitals offer a cybersecurity model [8] or highlight the problem from an organizational perspective [9], and a risk management angle [10].

To our knowledge, no paper has proposed to explore the cybersecurity environment of hospitals in order to validate cross working areas for both the academia and the practitioners. Cybersecurity frameworks and information security best practices are well established on the field. However, we have not found a study on the cybersecurity maturity level of healthcare institutions in general and hospitals in particular.

2 Research Questions

As healthcare systems continue to evolve, the need to comprehensively understand, address, and mitigate information security and privacy risks becomes imperative. Our project aims at addressing the following research questions:

- Research Question 1 (RQ1). What is the current state of the research on the application of information security and privacy in hospitals?
- Research Question 2 (RQ2). What are the differences between healthcare and social organisations based on the implementation of cybersecurity measures? And, what maturity model could emerge?
- Research Question 3 (RQ3). How to evaluate the maturity level for its usefulness in the context of hospitals?

3 Research Objectives

Our research objectives to perform this study are as follows.

- RQ1. To have a better understanding of previous studies by investigating the current state of research and to explore research gaps in the literature which could lead us to direct our PhD journey towards an end-result with an impact to the practitioners.
- RQ2. To develop a maturity level for healthcare and social organizations based on their implementation of cybersecurity best practices in order to assist managers in hospitals to brainstorm on their cybersecurity posture and facilitating cyber risks identification.

- RQ3. To evaluate the maturity model defined against tangible use-cases such as hospitals in order to attest of the usefulness of the tool (Fig. 1).

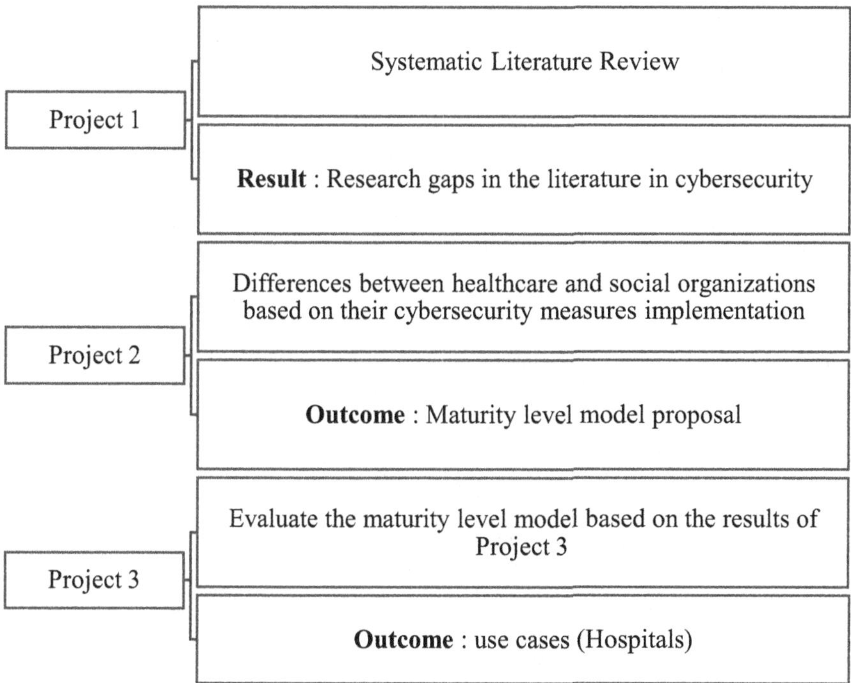

Fig. 1. Overview of Projects and expected outcomes

4 Proposed Research Methods

To respond to research questions, both qualitative and quantitative research methodologies will be used. Systematic Literature Review, interviews, case studies, and a large-scale survey methodology will be used mainly. We continue with proposing the research methods per RQ.

4.1 Methodology for RQ1

In RQ1, we used the systematic literature review which aims to present a comprehensive evaluation of a research topic [11] by following a structured methodology. For this purpose, SLRs require to understand the scope of the research by defining a clear objective. To review the application of information security and privacy in hospitals, we analysed the content and the metadata of the selected articles. We subdivided our main research question (RQ) in three detailed SLR-RQs to collect more knowledge on the subject.

We selected five academic databases (i.e., Web of Science (WoS), Scopus, AIS Electronic Library, Science Direct, IEEE Xplore Digital Library) because these databases are recognised for providing access to peer-reviewed publications in an intuitive and structured manner. We decided not to restrict the search to a specific period because the realm of cybersecurity coupled with the healthcare sector is rather new (Table 1).

Table 1. Research Gaps in the literature. Note: the results are not cumulative, because articles can be classified in more than one set of controls.

Research agenda	Example of topics	Number of papers with this recommendation
1) Big data	How to gather, process and analyse large volumes of personal healthcare data? How to build an efficient data management capability within hospitals? How to ensure anonymisation and encryption of patient data in the context of big data analysis?	16
2) Machine learning	How to optimize the healthcare system and provide intelligent services effectively? How to integrate machine learning and blockchain in the context of healthcare systems? What are the risks of AI and machine learning methods in hospitals?	23
3) Internet of Things (IOT)	What are the privacy issues related to medical devices (geolocation, monitoring, information sharing)? What are the security issues of medical devices' sensors and how to mitigate these risks? How IOT and 5G are related in the context of smart healthcare?	45
4) Cloud Computing	What are the risks and advantages of Cloud computing for patient data?	27
5) Blockchain	How blockchain can be used to enhance security and privacy risks of medical devices? How to improve blockchain stability for reliability purposes in the context of the healthcare?	32
6) Standards and regulations	How to vet new technologies related to medical devices from a security and privacy perspective? How to define a roadmap of an independent and trustworthy third party competent to audit hospitals' security and privacy compliance?	10

Therefore, all the results until May 2021 were considered, which is when our literature search ended. We searched for articles containing a combination of the following terms in their title, abstract and keywords: "cybersecurity AND health*"; "GDPR AND hospital"; "cybersecurity AND hospital*"; "information security* AND privacy AND health"; ""information privacy" AND hospital"; "information security AND privacy AND hospital". We excluded the duplicates emerging from the search of multiple databases and we proceeded to identify the articles' importance and relevance for our goal.

Results of RQ1

The result showed that the technical areas of cybersecurity were most tackled in the sample and less attention were paid to other realms such as the management, the policies, the processes, the culture. If the literature on cybersecurity in hospitals is increasing, it is still relatively limited compared to the level of threat faced by patient data. Also, we provided a research agenda by identifying in our sample key domains in information security and privacy where further research in hospitals is needed and we called for more research in integration of advanced methods; in novel architectures in order to ensure a comprehensive approach in handling information security and privacy in hospitals.

This research agenda highlighted research gaps in the literature, which was key in the decision making process to tackle our topic from an angle of "standards and regulations" and from a Design Science Research perspective.

4.2 Methodology for RQ2 (In Progress)

In the second study, we will focus on healthcare and social institutions classification based on their implementation of cybersecurity measures. We will derive key differentiators which could help in profiling these organizations and develop a maturity level model based on these unique differentiators.

The research will rely on a design science methodology to arise with various types of institutions based on their cybersecurity set of practices and the related differences. This phase is in progress and the statistical findings are derived based on a combination of a cluster analysis and an analysis of variance (Tukey Honest Significant Difference) to discover differences among the statistical clusters.

Outcome of RQ2

One of the current deliverables is to distinguish cybersecurity profiles of healthcare and social organizations as well as cybersecurity measures which can be considered as key differentiators in the profiling exercise, and investigate to what extent this could be linked to the performance outcome. This could constitute a first step exercise towards a cybersecurity maturity model which will help practitioners select security standards for hospitals. This third study intends to address a practical element of the research agenda formulated in RQ1 by providing to the industries an evaluation tool to flag cyber risks in hospitals (Table 2).

Table 2. Overview of profiles along differentiators by translating the feature analysis into a textual comparison.

Clusters	Cluster 1	Cluster 3	Cluster 2	Cluster 4	Cluster 5
Profiles	**Profile A**	**Profile B**	**Profile C**	**Profile D**	**Profile E**
Training	Lack of training	Moderate level of training	Moderate level of training	Severe lack of training	Superior level of training
Regular updates	Severe lack of regular updates	Severe lack of regular updates	Little or no regular updates	Moderate level of regular updates	High level of regular updates
Data backup	Lack of data backup	Lack of data backup	Moderate frequency of data backup	Frequent data backup	Moderate frequency of data backup

4.3 Methodology and outcome for RQ3

In our third project, an iterative validation of the maturity model and the associated maturity levels will constitute the objective [12]. We will use the UTAUT model for setting up an evaluation study according to an evaluation plan which is not yet defined. We will carry out an expert panel and a case study on hospitals to assess the suggested "model" artefact.

The purpose of the expert panel will be to redesign the cybersecurity maturity model that presents the relationships among the profiles constructs and showcase its impact on performance outcomes [13]. Additionally, we plan to ask the experts about their views to improve the exercise by providing more details on the cybersecurity factors which could play a key role in assessing the cybersecurity posture of a healthcare organization [14].

4.4 Research Limitation

Our study may be impacted by the limited dataset obtained to conduct the research. Furthermore, the development and evaluation of the cybersecurity models are still at the early stages and multiples parameters will need to be integrate in the research plan for effectiveness and consistency.

Besides, the evaluation of the maturity model will need to utilize several hospitals as use cases in order to have a tangible benchmark. Finally, the PhD overall project is a part-time PhD and therefore, the results, once obtained may no longer fit the reality of practitioners at the time of input collection (Table 3).

Table 3. Timeline

Timeline		
	Start Date	End date
Project I		
Data collection and database creation	January 2020	June 2020
Systematic Literature Review (SLR)	September 2021	May 2021
Analyzing results	June 2021	September 2021
Article Draft, peer-review, and consolidation	October 2021	January 2022
Project II		
Data cleaning	May 2022	September 2022
Statistical methods exploration	October 2022	June 2023
Statistical analysis validation	September 2023	January 2024
Application of the Design Science Research Methodology	February 2024	May 2024
Developing artefacts and build a maturity model	June 2024	September 2024
Article draft	September 2024	December 2024
Project III		
Literature Review	January 2025	March 2025
Exploration of methods to evaluate the maturity level	March 2025	May 2025
Use-case	June 2025	September 2025
Draft peer-review and article consolidation	September 2025	November 2025
Final Thesis	December 2025	March 2026
Final Defense	September 2026	

5 Conclusion and Future Steps

In conclusion, the doctoral consortium report has described three approaches to gradually understand the research field, identify gaps in the industry and facilitate the identification of cyber risks in hospitals. First, we described our observation of the literature in an SLR paper with a research gap in the literature as main result (RQ1). This allowed us to direct our overall research towards a design science research methodology. For this second project, we will classify healthcare and social institutions based on their implementation of cybersecurity measures in order to develop a cybersecurity maturity model unique for the sector (RQ2). The outcome will allow us to develop cybersecurity profiles for the healthcare sector, including hospitals and be able to evaluate the model against expert panels or a use case (RQ3).

We believe that the study will contribute to a better understanding of the state of hospitals in cybersecurity and facilitate cyber risks assessment thanks to the model built and evaluated.

Acknowledgments. This Ph.D. is organized by Ghent University (Belgium) under the supervision of Prof. Dr. Amy Van Looy and Prof. Dr. Geert Poels.

Disclosure of Interests. The authors have no competing interests to declare that are relevant to the content of this article.

References

1. Aarestrup, F.M., et al.: Towards a European health research and innovation cloud (HRIC). Genome Med. **12**(1) (2020). https://doi.org/10.1186/s13073-020-0713-z
2. Appari, A., Johnson, M.E.: Information security and privacy in healthcare: current state of research. Int. J. Internet Enterp. Manag. **6**(4), 279 (2010). https://doi.org/10.1504/IJIEM.2010.035624
3. Argaw, S.T., et al.: Cybersecurity of Hospitals: discussing the challenges and working towards mitigating the risks. BMC Med. Inform. Decis. Making **20**(1) (2020). https://doi.org/10.1186/S12911-020-01161-7
4. Becker, J., Knackstedt, R., Pöppelbuß, J.: Developing maturity models for IT management. Bus. Inf. Syst. Eng. **1**(3), 213–222 (2009). https://doi.org/10.1007/S12599-009-0044-5
5. Fernández-Alemán, J.L., Señor, I.C., Lozoya, P., Ángel, O., Toval, A.: Security and privacy in electronic health records: a systematic literature review. J. Biomed. Inform. **46**(3), 541–562 (2013). https://doi.org/10.1016/J.JBI.2012.12.003
6. Gollhardt, T., Halsbenning, S., Hermann, A., Karsakova, A., Becker, J.: Development of a digital transformation maturity model for IT companies. In: Proceedings - 2020 IEEE 22nd Conference on Business Informatics, CBI 2020, vol. 1, pp. 94–103 (2020). https://doi.org/10.1109/CBI49978.2020.00018
7. Jalali, M.S., Kaiser, J.P.: Cybersecurity in hospitals: a systematic, organizational perspective. J. Med. Internet Res. **20**(5) (2018). https://doi.org/10.2196/10059
8. Jofre, M., et al.: Cybersecurity and privacy risk assessment of point-of-care systems in healthcare—a use case approach. Appl. Sci. **11**(15), 6699 (2021). https://doi.org/10.3390/APP11156699
9. Keele, S.: Guidelines for performing systematic literature reviews in software engineering. Technical report, Ver. 2.3 EBSE Technical Report. EBSE (2007)
10. Martin, G., Martin, P., Hankin, C., Darzi, A., Kinross, J.: Cybersecurity and healthcare: How safe are we? BMJ **358** (2017). https://doi.org/10.1136/bmj.j3179
11. Mettler, T.: Maturity assessment models: a design science research approach. Int. J. Soc. Syst. Sci. **3**(1/2), 81 (2011). https://doi.org/10.1504/IJSSS.2011.038934
12. Muthuppalaniappan, M., Stevenson, K.: Healthcare cyber-attacks and the COVID-19 pandemic: an urgent threat to global health. Int. J. Qual. Health Care **33**(1) (2021). https://doi.org/10.1093/INTQHC/MZAA117
13. Naconha, A.E.: A Cybersecurity Model for the Health Sector: A Case Study of Hospitals in Nairobi, Kenya, vol. 4, no. 1, p. 6 (2021). http://erepo.usiu.ac.ke/11732/6742
14. Zafar, H., Ko, M.S., Clark, J.G.: Security risk management in healthcare: a case study. Commun. Assoc. Inf. Syst. **34**(1), 737–750 (2014). https://doi.org/10.17705/1cais.03437

Towards a Hybrid Intelligence Paradigm: Systematic Integration of Human and Artificial Capabilities

Antoni Mestre[1,2](✉)

[1] VRAIN Institute, Universitat Politècnica de Valencia, Camino de Vera, S/N, 46022 Valencia, Valencia, Spain
anmesgas@vrain.upv.es

[2] valgrAI (Valencian Graduate School and Research Network of Artificial Intelligence), Valencia, Spain

Abstract. The evolution of Artificial Intelligence from traditional inference-based systems to sophisticated generative models has blurred the boundaries between machine and human capabilities, giving rise to Hybrid Intelligence (HI). HI represents a symbiotic relationship between human and artificial intelligence, integrating human wisdom and expertise with machine intelligence. This work aspires to explore the paradigm shift towards HI, with a focus on integrating human expertise with machine intelligence. It aims to address challenges in human-machine interaction and dynamic task management within HI systems, emphasizing the necessity for seamless integration to fully exploit the capabilities of both entities. Through interdisciplinary collaboration and empirical inquiry, this research endeavors to advance understanding and implementation of HI systems across diverse domains, paving the way for systems that harness the intelligence of humans and machines to tackle complex challenges.

Keywords: Hybrid intelligence · human-machine interaction · human-centered artificial intelligence · dynamic task management

1 Introduction

Artificial Intelligence (AI) has undergone a remarkable evolution in recent years, transitioning from traditional inference-based systems to sophisticated generative models [1]. This progression has blurred the boundaries between machine and human capabilities significantly, marking the emergence of a novel paradigm known as Hybrid Intelligence (HI) [2]. HI refers to a symbiotic relationship between human and artificial intelligence, where their unique strengths converge to achieve superior collective intelligence beyond the capabilities of either entity alone [3].

The transition towards HI has been promoted by advances in machine learning algorithms, natural language processing, and machines' ability to understand complex contexts [4]. This shift has not only enabled machines to understand human language more naturally but also to generate creative content. Generative models have fostered AI

J. Araújo et al. (Eds.): RCIS 2024, LNBIP 514, pp. 149–156, 2024.
https://doi.org/10.1007/978-3-031-59468-7_17

beyond mere pattern reproduction, enabling the creation of new and original content. Consequently, the distinction between artificial and human intelligence has become increasingly ambiguous [5].

HI does not seek to mimic human cognitive abilities but rather integrates human wisdom and intuition into the fabric of environments to coexist and cooperate with intelligent systems, as sowed in Fig. 1.

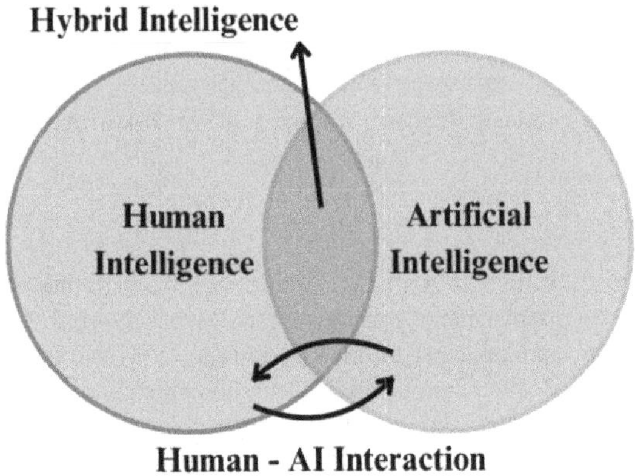

Fig. 1. Human, artificial and hybrid intelligence [5]

In this context of convergence, the integration between humans and machines must transcend static and predefined interactions. With AI now capable of dealing with complex contexts and adapting to dynamic scenarios, there's a need for a more fluid and organic interaction. Achieving real integration between humans and machines is imperative to fully leverage the capabilities of both. This paradigm shift requires a redefinition of how we conceptualize and experience collaboration between humans and intelligent systems. Despite the increasingly blurred line between AI and human capabilities, AI systems often lack the comprehensive understanding, adaptability, and collaborative prowess inherent in humans [6]. Conversely, humans, equipped with unique faculties such as flexible reasoning and the ability to adapt to sudden changes, can significantly benefit from the processing and learning capabilities of machines. The synergistic collaboration between these two entities involves overcoming complex challenges in task management to achieve shared goals [7].

This doctoral research initiative addresses the evolving landscape of human-machine collaboration within the framework of HI. We seek to contribute to the advancement of understanding and implementation of HI across diverse domains. Central to our work is the exploration of the intricate interplay between human cognition and artificial intelligence, emphasizing the need for seamless integration and synergy between these actors. By addressing these challenges, our aim is to pave the way for the realization of HI systems that leverage the collective intelligence of humans and machines to tackle complex problems. In this evolving landscape, such systems drive innovation across a spectrum of domains, and a comprehensive understanding of their dynamics is essential for the continued advancement of Information Science.

This doctoral paper is structured as follows: Sect. 2 outlines the motivation behind the project, focusing on addressing the challenges of human-machine interaction and dynamic task management in HI. It also sets out the research goals and key questions. Section 3 describes the interdisciplinary research methodology to tackle these questions. Finally, Sect. 4 summarizes the proposal and outlines future plans.

2 Research Objectives

This Ph.D. research is focused on exploring and improving the symbiotic relationship between the two main actors of HI; human and intelligent systems. By analyzing recent advancements and key concepts in human-machine interaction and the role assignment, we aim to uncover the intricate dynamics of IH.

In this section, we provide the groundwork for this doctoral research by outlining the fundamental principles and challenges that motivate our investigation, alongside presenting the research questions.

2.1 Problem Statement

The effective integration of human expertise with machine intelligence in HI systems hinges upon the development of novel collaborative frameworks and interaction methods that enable software engineers to achieve seamless and intuitive collaboration. Recent research has made significant strides in this direction, exploring various approaches to facilitate natural, efficient, and non-intrusive communication between humans and machines.

One prominent line of inquiry revolves around human-centered design [8], aiming to tailor HI systems to human needs, capabilities, and preferences [9]. Studies have emphasized the importance of understanding user requirements and ensuring that HI systems are intuitive and easy to use, thus promoting user acceptance and adoption [10].

Furthermore, advances in natural language processing [11] and affective computing have enabled HI systems to interpret human emotions [12] and intentions [13], leading to more empathetic and responsive interactions [14]. By incorporating affective cues into human-machine communication, researchers have sought to enhance the user experience and foster deeper engagement with HI systems [15].

However, despite these advancements, challenges remain in achieving seamless integration and understanding between humans and machines [16]. Current HI systems often struggle to adapt to dynamic user preferences and contexts, leading to suboptimal interaction experiences [15]. Moreover, ensuring the privacy and security of user data in HI environments poses significant ethical and technical challenges that require further research [17].

The efficient distribution of roles between actors involved in collaborative environments, based on their capabilities and real-time context, is a pivotal aspect of HI systems [18]. Dynamic task management is a critical aspect of HI systems [8], enabling the efficient distribution of tasks between human and machine actors based on their capabilities and real-time context. Recent research [19, 20] has explored various strategies and methodologies for achieving production-based optimal task management in HI environments.

One notable area of research focuses on adaptive task allocation algorithms that dynamically adjust task assignments based on changing environmental conditions and user preferences [21, 22]. By leveraging machine learning and optimization techniques, researchers have developed algorithms capable of optimizing task allocation in real time, thereby improving system performance and user satisfaction.

Additionally, studies have highlighted the importance of considering human factors such as cognitive load and fatigue when allocating tasks in HI systems [23]. By modeling human cognitive processes and physiological responses [24], researchers aim to develop task allocation strategies that minimize user stress and maximize productivity.

Despite these advancements, several challenges persist in the field of dynamic task allocation in HI systems [25]. Current approaches often struggle to balance task efficiency with user well-being, leading to suboptimal outcomes in terms of both performance and user satisfaction [26]. Moreover, the scalability and generalizability of existing task allocation algorithms remain limited mainly to assembly scenario [23, 26], hindering their applicability to diverse HI domains.

Addressing these challenges is paramount to pushing the boundaries of HI and amplifying its practical utility across diverse domains. Collaborative efforts across interdisciplinary teams will be crucial in overcoming these hurdles and realizing the full potential of HI systems in real-world settings.

2.2 Research Questions

With this aim in mind, the following Research Questions (RQ) emerge as pillars upon which this investigation stands:

RQ1: How can novel collaborative frameworks be developed to effectively integrate human expertise with machine intelligence in hybrid intelligence systems?

This question propels our inquiry into the realm of collaborative frameworks, aiming to develop tools such as taxonomies or ontologies to provide an organized and semantic structure that enhances interoperability, information retrieval, and facilitates the design of HI systems. By leveraging these approaches, we seek to establish comprehensive frameworks that seamlessly integrate human expertise and machine intelligence, fostering synergistic relationships.

RQ2: What human-centered strategies and interaction methods can be employed to achieve genuine, efficient, and non-intrusive communication in hybrid intelligence systems?

Delving into the intricacies of communication, this question probes the landscape of interaction strategies and methods within HI systems. With a steadfast commitment to human-centered design principles, our exploration aims to cultivate efficient and non-intrusive communication strategies, thereby nurturing trust, enhancing user experience, and facilitating seamless collaboration between human users and intelligent systems.

RQ3: What are the relevant criteria for the efficient and human-centered management of tasks in hybrid intelligence environments?

This question directs our attention to the domain of task management within HI environments, prompting an examination of criteria and guidelines essential for efficiency and human-centeredness. By delineating key factors that underpin effective task management, we endeavor to optimize system performance, bolster user satisfaction, and safeguard the well-being of human users.

RQ4: What methods and tools can be implemented for dynamic and human-centered task management in HI systems, ensuring optimal performance and adaptability to changing contexts?

Through the integration of flexible management mechanisms and adaptive AI-based technologies, our endeavor aims to optimize performance and ensure adaptability, all while maintaining a human-centered focus on collaboration between human users and intelligent systems.

3 Research Approach

In this doctoral research, the Design Science Research (DSR) [27] methodology is adopted to systematically address the challenges inherent in the development of the RQ exposed in the previous section. This methodology emphasizes the iterative creation and evaluation of innovative artifacts to effectively solve real-world problems.

To address **RQ1**, we will conduct a comprehensive literature review to gather insights from existing frameworks that address human-machine interrelationships. These insights may help to enhance our understanding of human-machine integration within the framework of HI. Subsequently, interdisciplinary collaborations with experts from fields such as software development and cognitive psychology will shape the design of a framework. Through this process, we aim to establish a structured framework that delineates the interrelationships between human expertise and machine intelligence. This framework will serve as a foundational tool for facilitating communication, collaboration, and decision-making within HI systems. By proposing an ontology, we can effectively capture and represent the complex interdependencies and interactions between human and machine actors, thus enabling the seamless integration of their capabilities.

Moving on to **RQ2**, we will examine human-centered interaction methods in HI systems through a review of empirical studies and user interviews, aiming to gain a comprehensive understanding of user preferences and model them accordingly. Drawing from these insights, we will develop interaction methods that align with human-centered design principles. Usability testing will be carried out using simulators of HI systems, such as smart homes or smart industrial setups, to guide iterative improvements aimed at enhancing user experience and engagement.

Addressing **RQ3** will begin with a comprehensive literature review to identify existing frameworks, models, and guidelines relevant to task management in hybrid intelligence environments. This review aims to delineate key factors and criteria crucial for efficient and human-centered task management. Following this, we will conduct focus group discussions and expert interviews involving professionals from diverse fields to gather insights and potentially uncover additional criteria not explicitly addressed in existing literature. Synthesizing findings from the literature review and discussions, we will formulate a preliminary set of criteria, encompassing aspects such as workload distribution, human emotional states, and adaptability to dynamic environments. Subsequently, we will administer surveys to validate the proposed criteria and prioritize them based on the perceived importance and feasibility of implementation among end-users and stakeholders.

Finally, for **RQ4** we will proceed to develop prototypes by integrating methods and tools that provide dynamic task management functionalities considering the criteria identified in RQ3. These prototypes will undergo testing within simulated environments of HI systems. Prototypes will be instrumental in testing and refining the proposed methods and tools. Subsequently, we will collect data from the usage of these prototypes. Statistical analysis techniques will then be employed to identify correlations and patterns in the collected data, facilitating the development of machine-learning-based predictive models. These models will undergo validation through empirical testing and further refinement based on feedback and additional data collection.

In the future, our research outcomes could be used for the development of a HI system capable of aligning with users' preferences, optimizing efficiency, and effectiveness while maintaining a human-centric approach. Currently, our research is in its early planning stage, and data collection and analysis have not yet commenced. As our methodology continues to evolve, we are committed to adhering to open science principles, ensuring the accessibility and reproducibility of our data and tools. Furthermore, we aim to present our work at academic forums, contributing to the broader research community and facilitating the dissemination of our findings. It should be noted that this work has not been presented at other doctoral consortia or conferences.

4 Conclusion

In this doctoral consortium paper, we have presented an overview of our research agenda focused on advancing the paradigm of HI by integrating human and artificial capabilities synergistically. Our investigation delves into the intricacies of human-machine interaction and dynamic task management, aiming to address key challenges and propel the field forward.

Through interdisciplinary collaboration and rigorous empirical inquiry, we seek to address fundamental research questions concerning human-machine relational frameworks, human-centered interaction methods, criteria for efficient task management, and task management methodologies in HI systems. By systematically investigating these questions, we aim to lay the groundwork for the development of HI systems that leverage the intelligence of humans and machines to tackle complex problems across diverse domains.

By addressing the challenges and complexities inherent in this domain, we strive to drive innovation and advancement in the field of information science, ultimately benefiting society as a whole.

Acknowledgments. I would like to express my gratitude to the supervisors of this Ph.D., Dr. Manoli Albert and Prof. Dr. Vicente Pelechano, from the Polytechnic University of Valencia (Spain) for their invaluable guidance and support. Additionally, I extend my sincere appreciation to Dr. Miriam Gil from the University of Valencia (Spain) for her valuable insights and advice. Furthermore, I acknowledge the financial support provided by the Generalitat Valenciana under project GV/2021/072 and by the MINECO under project AVANTIA PID2020-114480RB-I00.

References

1. Rayhan, S., Rayhan, A.: AI Odyssey: Unraveling the Past, Mastering the Present, and Charting the Future of Artificial Intelligence (2023)
2. Dellermann, D., Ebel, P., Söllner, M., Leimeister, J.M.: Hybrid intelligence. Bus. Inf. Syst. Eng. **61**(5), 637–643 (2019). https://doi.org/10.1007/s12599-019-00595-2
3. Akata, Z., et al.: A research agenda for hybrid intelligence: augmenting human intellect with collaborative, adaptive, responsible, and explainable artificial intelligence. Computer **53**(8), 18–28 (2020). https://doi.org/10.1109/MC.2020.2996587
4. Rane, N.: ChatGPT and Similar Generative Artificial Intelligence (AI) for Smart Industry: Role, Challenges and Opportunities for Industry 4.0, Industry 5.0 and Society 5.0. Rochester, NY, 31 May 2023. https://doi.org/10.2139/ssrn.4603234
5. Jarrahi, M.H., Lutz, C., Newlands, G.: Artificial intelligence, human intelligence and hybrid intelligence based on mutual augmentation. Big Data Soc. **9**(2), 20539517221142824 (2022). https://doi.org/10.1177/20539517221142824
6. Zhou, L., et al.: Intelligence augmentation: towards building human-machine symbiotic relationship. AIS Trans. Hum.-Comput. Interact. **13**(2), 243–264 (2021). https://doi.org/10.17705/1thci.00149
7. Chen, A., Xiang, M., Wang, M., Lu, Y.: Harmony in intelligent hybrid teams: the influence of the intellectual ability of artificial intelligence on human members' reactions. Inf. Technol. People **36**(7), 2826–2846 (2022). https://doi.org/10.1108/ITP-01-2022-0059
8. Kadir, B.A., Broberg, O.: Human-centered design of work systems in the transition to industry 4.0. Appl. Ergon. **92**, 103334 (2021). https://doi.org/10.1016/j.apergo.2020.103334
9. Rožanec, J.M., et al.: Human-centric artificial intelligence architecture for industry 5.0 applications. Int. J. Prod. Res. **61**(20), 6847–6872 (2023). https://doi.org/10.1080/00207543.2022.2138611
10. Margetis, G., Ntoa, S., Antona, M., Stephanidis, C.: Human-centered design of artificial intelligence. In: Salvendy, G., Karwowski, W. (eds.) Handbook of Human Factors and Ergonomics, 1st edn, pp. 1085–1106. Wiley (2021). https://doi.org/10.1002/9781119636113.ch42
11. McIntosh, T.R., Susnjak, T., Liu, T., Watters, P., Halgamuge, M.N.: From Google Gemini to OpenAI Q* (Q-Star): A Survey of Reshaping the Generative Artificial Intelligence (AI) Research Landscape. arXiv (2023). http://arxiv.org/abs/2312.10868. Accessed 08 Jan 2024
12. Khare, S.K., Blanes-Vidal, V., Nadimi, E.S., Acharya, U.R.: Emotion recognition and artificial intelligence: a systematic review (2014–2023) and research recommendations. Inf. Fusion **102**, 102019 (2024). https://doi.org/10.1016/j.inffus.2023.102019
13. Konar, A.: Artificial Intelligence and Soft Computing: Behavioral and Cognitive Modeling of the Human Brain. CRC Press (2018)
14. Yalçın, Ö.N.: Evaluating empathy in artificial agents. In: 2019 8th International Conference on Affective Computing and Intelligent Interaction (ACII), pp. 1–7 (2019). https://doi.org/10.1109/ACII.2019.8925498

15. Gurcan, F., Cagiltay, N.E., Cagiltay, K.: Mapping human-computer interaction research themes and trends from its existence to today: a topic modeling-based review of past 60 years. Int. J. Hum.-Comput. Interact. **37**(3), 267–280 (2021). https://doi.org/10.1080/10447318.2020.1819668
16. Mueller, F.F., et al.: Next steps for human-computer integration. In: Proceedings of the 2020 CHI Conference on Human Factors in Computing Systems, in CHI 2020, pp. 1–15. Association for Computing Machinery, New York (2020). https://doi.org/10.1145/3313831.3376242
17. Safdar, N.M., Banja, J.D., Meltzer, C.C.: Ethical considerations in artificial intelligence. Eur. J. Radiol. **122**, 108768 (2020). https://doi.org/10.1016/j.ejrad.2019.108768
18. Pescetelli, N.: A brief taxonomy of hybrid intelligence. Forecasting **3**(3), Art. no. 3 (2021). https://doi.org/10.3390/forecast3030039
19. Yu, T., Huang, J., Chang, Q.: Optimizing task scheduling in human-robot collaboration with deep multi-agent reinforcement learning. J. Manuf. Syst. **60**, 487–499 (2021). https://doi.org/10.1016/j.jmsy.2021.07.015
20. Li, S., Wang, R., Zheng, P., Wang, L.: Towards proactive human–robot collaboration: a foreseeable cognitive manufacturing paradigm. J. Manuf. Syst. **60**, 547–552 (2021). https://doi.org/10.1016/j.jmsy.2021.07.017
21. Pupa, A., Van Dijk, W., Secchi, C.: A human-centered dynamic scheduling architecture for collaborative application. IEEE Robot. Autom. Lett. **6**(3), 4736–4743 (2021). https://doi.org/10.1109/LRA.2021.3068888
22. Süße, T., Kobert, M., Kries, C.: Human-AI interaction in remanufacturing: exploring shop floor workers' behavioural patterns within a specific human-AI system. Labour Ind. 1–20 (2023). https://doi.org/10.1080/10301763.2023.2251103
23. Yao, B., Li, X., Ji, Z., Xiao, K., Xu, W.: Task reallocation of human-robot collaborative production workshop based on a dynamic human fatigue model. Comput. Ind. Eng. 109855 (2023). https://doi.org/10.1016/j.cie.2023.109855
24. Pimenta, A., Carneiro, D., Neves, J., Novais, P.: A neural network to classify fatigue from human–computer interaction. Neurocomputing **172**, 413–426 (2016). https://doi.org/10.1016/j.neucom.2015.03.105
25. Kiyokawa, T., et al.: Difficulty and complexity definitions for assembly task allocation and assignment in human–robot collaborations: a review. Robot. Comput.-Integr. Manuf. **84**, 102598 (2023). https://doi.org/10.1016/j.rcim.2023.102598
26. Lee, M.-L., Behdad, S., Liang, X., Zheng, M.: Task allocation and planning for product disassembly with human–robot collaboration. Robot. Comput.-Integr. Manuf. **76**, 102306 (2022). https://doi.org/10.1016/j.rcim.2021.102306
27. Johannesson, P., Perjons, E.: An Introduction to Design Science. Springer, Cham (2014). https://doi.org/10.1007/978-3-319-10632-8

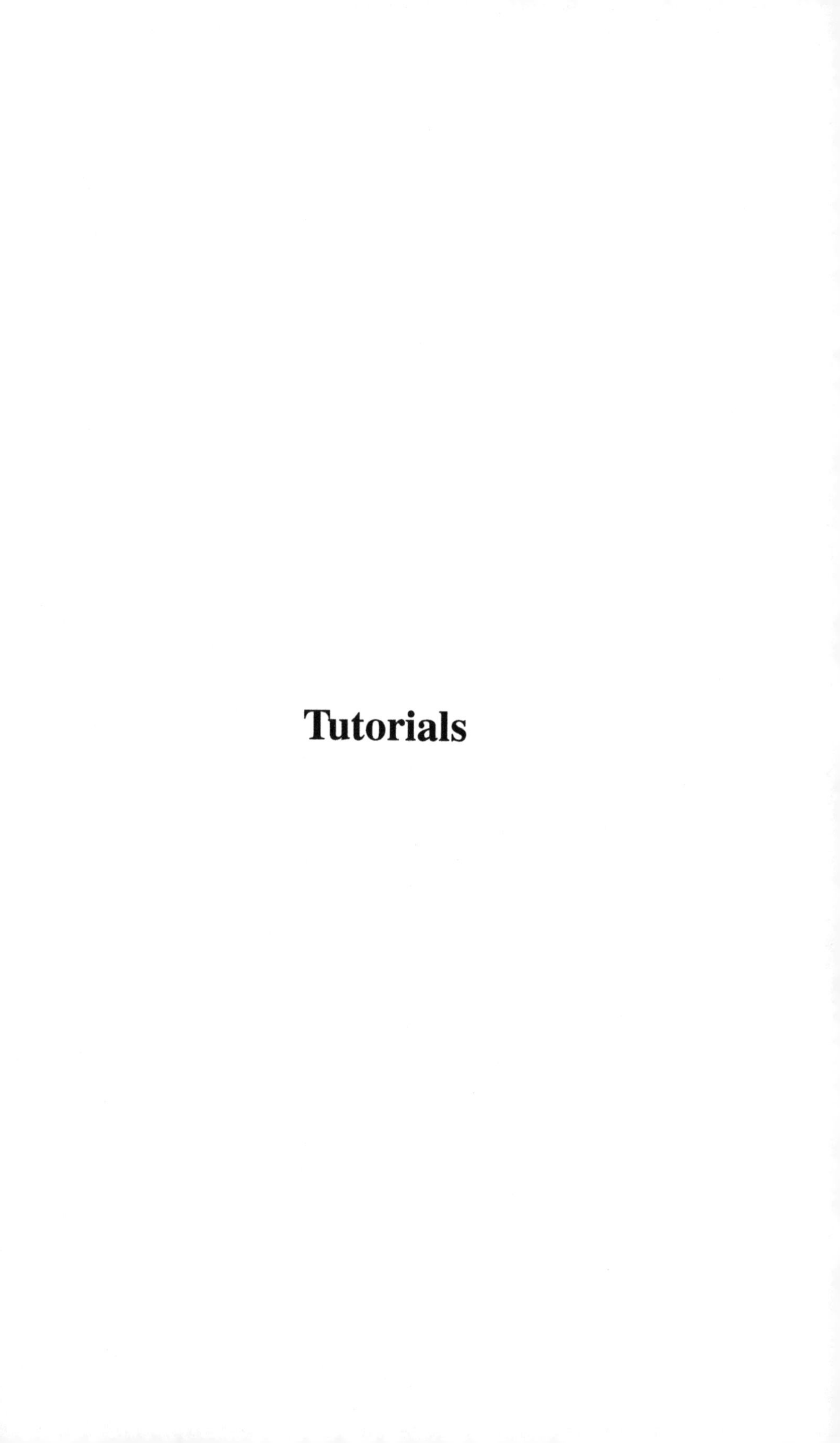

Tutorials

How Data Analytics and Design Science Fit: A Joint Research-Methodological Perspective

Faiza A. Bukhsh and Maya Daneva(✉)

University of Twente, Enschede, The Netherlands
{f.a.bukhsh,m.daneva}@utwente.nl

Abstract. Design Science Research (DSR) is about the design of artefacts and their study in context. DSR is widely used in the field of Information Systems Research and, in the past decade, also in Empirical Software Engineering. In a research project, Design Science is used to first identify the type of problem at hand and then follow the right steps to solve it. However, Data Analytics projects have a different focus and foundation, compared to information systems/software development projects. Whereas in the latter, the focus is on the product, in Data Analytics the focus is on the data, preparation, and interpretation.

In this tutorial, we explore the possible ways of applying DSR methodologies to Data Analytics research projects. To this end, we first examine existing data science methodologies from a DSR perspective. Second, selected DSR approaches are examined from a data science perspective. Data science-focused factors such as data preparation, analytics, interpretation, privacy and biases are discussed as part of research-methodological steps.

Keywords: Industry-relevant research · empirical research methods · empirical evaluation · data science · data analytics · generalizability

1 Introduction

Design Science Research (DSR) is a research paradigm widely adopted in the research communities of Information Systems (IS) and Empirical Software Engineering (ESE). Research methodologists define DSR as the methodology that seeks to "invent (i.e. design) new means for acting in the world in order to change and improve reality" (e.g. [1]). In turn, DSR influences reality through creating and evaluating artefacts that serve stakeholders' purposes and solve stakeholders' problems [2]. Within this paradigm, a number of approaches were developed (e.g. [2, 3]). In each approach, when employed in IS or ESE research projects, the researcher first identifies the type of problem at hand, and then applies a methodologically sound series of steps to solve it. However, the recent proliferation of data science methodologies and the specific problems pertaining to their creation and evaluation, posed a challenge to many researches regarding the selection and application of DSR approaches to data science research contexts. This is because Data Analytics/Data Science projects have a different focus and foundation, compared to software/systems engineering projects. Whereas, in the latter, the focus is

J. Araújo et al. (Eds.): RCIS 2024, LNBIP 514, pp. 159–160, 2024.
https://doi.org/10.1007/978-3-031-59468-7

on the product, in Data Analytics, all the discussions revolve around the data, preparation, and interpretation.

This tutorial introduces a joint perspective on data science methodologies and DSR methodologies. The goal of the tutorial is twofold: (1) to rethink existing data science methodologies from a DSR perspective, and (2) to examine existing DSR methodologies—such as those of Wierienga [2] and of Peffers et al. [3], from a data science perspective. The overall objective is to help the tutorial attendees to evaluate the fit between data science research contexts and the available DSR methodologies and select the research methodology to be used in that research context. At the end of the tutorial, attendees should be able to critically reason about the possible choices in designing a design-science-grounded research process and the quality criteria for evaluating their artefacts, in a data science (also called Data Analytics) research project. To achieve this, the tutorial focuses on two Data Analytics research contexts (namely CRISP-DM [4]) and SEMMA [5]) and two popular DSR methodologies ([2] and [3]). For each Data Analytics context, the type of problems is explored, the characteristics of the projects are examined, and based on this, a recommendation for a DSR methodology is made and validity questions [6] are discussed.

The tutorial content is organized in three parts. Part I presents the overall layout of the process inspired by the DSR paradigm and its particular applications according to Wieringa [2] and to Peffers et al. [3]. Special attention is paid on the design choices that researchers face in data science projects. Part II presents the Data Analytics research contexts related to SEMMA [5] and CRISP-DM [4]. This part draws on the first author's own experience in teaching SEMMA and CRISP-DM to PhD students in the Netherlands' National PhD School SIKS. Part III is interactive and goes deeper on how Data Analytics benefits from the application of DSR. In each of the two Data Analytics contexts, data science-focused factors such as data preparation, analytics, interpretation, privacy and biases are discussed as part of research-methodological steps. Part III concludes with discussion on observations in industry practice of data analytics, and with reflection on validity questions and reflection on generalizability [6].

References

1. Venable, J.R., Pries-Heje, J., Baskerville, R.: Choosing a design science research methodology. In: ACIS2017 Conference Proceeding University of Tasmania (2017)
2. Wieringa, R.: Design Science Methodology for Information Systems and Software Engineering. Springer, Heidelberg (2014). https://doi.org/10.1007/978-3-662-43839-8
3. Peffers, K., Tuunanen, T., Rothenberger, M.A., Chatterjee, S.: A design science research methodology for information systems research. J. Manag. Inf. Syst. **24**(3), 45–77 (2008)
4. Chapman, P., Clinton, J., Kerber, R., Khabaza, T., Reinartz, T., Shearer, C., Wirth R.: The CRISP-DM User Guide (2000)
5. Azevedo, A., Santos, M.F.: KDD, SEMMA and CRISP-DM: a parallel overview. In: Proceedings of the IADIS European Conference on Data Mining, pp. 182–185 (2008)
6. Wieringa, R.J., Daneva, M.: Six strategies for generalizing software engineering theories. Sci. Comput. Program. **101**, 136–152 (2015)

Decision Intelligence for Enterprise and IS Engineering

Jan Vanthienen(✉)

Department of Decision Sciences and Information Management (LIRIS), KU Leuven, Leuven, Belgium
jan.vanthienen@kuleuven.be

Abstract. Decisions are everywhere. Modelling decisions is important in processes, information systems, service applications, analytics, and so many other areas. Modelling decisions in the correct way is vital. This tutorial is about decision modelling, the new decision model and notation (DMN) standard, and the role of decisions as an integral part of information systems engineering.

The tutorial introduces decision modelling and provides an overview of three typical research areas: (1) Building decision models from text or event logs, using deep learning, decision mining or other techniques; (2) Integrating decision and process models through separation of concerns; and (3) Using intelligent explanation services with a generic decision model chatbot.

Keywords: decision modelling · DMN · explainable decisions

1 Introduction

Knowledge-intensive processes incorporate lots of decisions and decision knowledge. Decision models offer unique features to capture the business logic of those decisions effectively and correctly by the business. Since the introduction of the Decision Model and Notation (DMN) standard [1], decision management and modelling have become an important research subject and are increasingly being used in industry.

2 Important Research Areas

The primary goal of DMN is to provide a common notation, and to bridge the gap between business decision design and implementation. DMN provides distinct, but related constructs for decision modelling: the decision requirements diagram, the decision logic, and the corresponding expression language, called FEEL [2]. This section indicates three contemporary research topics in decision modelling and execution.

J. Araújo et al. (Eds.): RCIS 2024, LNBIP 514, pp. 161–163, 2024.
https://doi.org/10.1007/978-3-031-59468-7

2.1 Decisions and Processes

Modelling business processes is essential for business effectiveness and efficiency. But knowledge-intensive processes incorporate lots of decision knowledge, that should not be hidden in process flows, because hardcoding (decision) rules in processes leads to complex and inflexible process models. In analogy with the Business Process Modelling & Notation Standard (BPMN), the Decision Model & Notation standard (DMN) was developed, allowing to model decisions and processes separately. Decisions should not be considered as local or unrelated to other elements of the process. Decision models can contain multiple related decisions in a single decision model, with elements in common (e.g. decision logic, input data). These decisions, however, may extend over process modelling elements, produce intermediate events or data, or require a specific ordering in the process model, so the decision model is not completely isolated.

2.2 Building Decision Models Automatically

Next to manually building models by domain experts, new AI techniques might allow extracting a decision model from textual descriptions in a (semi-)automatic way using natural language processing. Different ways have been explored: (1) Model construction based on language patterns for dependencies, and (2) Automatic construction using deep learning techniques (BERT, GPT) [3]. These approaches using deep learning or natural language processing offer interesting research challenges.

2.3 Using Decision Models for Intelligent Explanation

Despite the digital transformation and automation of day-to-day processes, end users still have many questions regarding applications, procedures or decisions. Organizations struggle with these requests and redirect them to call centers, service desks, FAQs, and conversational chatbots. Sadly, these expensive initiatives do not always meet customer demands because of long waiting times, incorrect advice, liability and/or complex human interventions. Decision modelling presents an interesting new research path. The decision model approach uses DMN models not just to make a decision in a knowledge-intensive process, but also as a knowledge base to enable reasoning and automatically create intelligent assistance for various decision related questions. Research studies the explanation quality of such an intelligent assistant, compared to manual explanations and ChatGPT [4].

References

1. Object Management Group: Decision Model and Notation, Version 1.3 (2019). https://www.omg.org/spec/DMN/1.3/PDF
2. Vanthienen, J.: Decisions, advice and explanation: an overview and research agenda. In: Liebowitz, J. (ed.) A Research Agenda for Knowledge Management and Analytics, p. 256. Edward Elgar Publishing (2021)

3. Goossens, A., De Smedt, J., Vanthienen, J.: Extracting decision model and notation models from text using deep learning techniques. Expert Syst. Appl. **211**, 118667 (2023). https://doi.org/10.1016/j.eswa.2022.118667
4. Goossens, A., Vanthienen, J.: From text to intelligent services in knowledge intensive decision processes: Text2Chat. In: Proceedings of the 57th Hawaii International Conference on System Sciences (HICSS), Hawaii (2024)

Comparing Products Using Similarity Matching

Mike Mannion[1(✉)] and Hermann Kaindl[2]

[1] Glasgow Caledonian University, Glasgow G4 7BA, UK
m.a.g.mannion@gcu.ac.uk
[2] Technical University of Vienna, Vienna, Austria

Abstract. This paper describes a tutorial that shows an approach to product comparison using similarity matching in which product features, configured from a software product line feature model, are represented by a binary string and the comparison achieved using a binary string metric.

Keywords: Product Comparison · Feature Similarity · Binary String

1 Introduction

A software product line is a set of products that share a set of software features and assets that satisfy the specific needs of one or more target markets. The discipline of systematically planning, constructing, evolving and managing that set of products is known as Software Product Line Engineering (SPLE). In SPLE, a feature model is often constructed as a hierarchy of features with some additional cross-cutting relationships. Such feature models facilitate a certain kind of reuse.

As the volume, variety and velocity of products in software-intensive systems product lines increase, product comparison is difficult when each product has hundreds of features. Reasons for product comparison include (i) concern for sustainability reasons whether to build a new product or not (ii) evaluating how products differ for strategic positioning reasons (iii) gauging if a product line needs to be reorganized (iv) assessing if a product falls within legislative and regulatory boundaries.

We will describe a product comparison approach using similarity matching. A product configured from a product line feature model is represented as a weighted binary string where 1 represents a feature's presence, 0 represents its absence, and a weight represents a feature's relative importance to the product.

Relative importance values influence similarity matching so that the features considered important are the ones that primarily influence what is judged to be similar. The similarity between products is compared using a binary string metric. For a product line that contains thousands of features the allocation of relative importance values is only practical when done automatically. However, methods for the allocation of feature weights are contested. Feature models are often represented as trees in which a feature is a node in the tree and a relationship between features is an edge. A feature's relative importance is calculated as a function of different structural measures of the tree. We

J. Araújo et al. (Eds.): RCIS 2024, LNBIP 514, pp. 164–165, 2024.
https://doi.org/10.1007/978-3-031-59468-7

Table 1. Tutorial Structure

Part 1: Trends in Software Product Line Development Software Product Line Development; Demand Trends – Personalization; Supply Trends – Customization; Matching Supply & Demand – Need for Product Comparisons
Part 2: Reuse Approaches - Feature Model-Based Development Feature Models; Product Derivation; Selection from a Product Line Model of Features; Product Verification: Constraint-based Satisfaction, Examples; Tools.
Part 3: Product Comparison using Similarity Matching Techniques Product and Feature Comparisons; Similarity in Domain & Application Engineering; The Role of Feature-Similarity Models; A Feature-Similarity Model Construction Process.
Part 4: Using Binary Strings for Feature Similarity Matching Representing Product Configurations with Binary Strings; Binary String Metrics; Positive vs. Negative Matches; Weighted Binary Strings; Allocating Weights, Examples; Open-ended Research Questions.
Part 5: Summary Lessons Learned, Future Challenges, Recap, Reflections

will describe one weight allocation method using a combination of such measures and discuss the benefits and limitations of this method using a mobile phone example.

2 Topics

Table 1 shows the coverage of the topics and the structure of the tutorial.

Disclosure of Interests. The authors have no competing interests to declare that are relevant to the content of this article.

References

1. Mannion, M., Kaindl, H.: Determining the relative importance of features for influencing product similarity matching. In: Proceedings of the IEEE 47th Annual Computers, Software, and Applications Conference (COMPSAC), Torino, Italy, 26–30 June, 2023, pp. 1638–1645 (2023). ISBN 979-8-3503-2697-0
2. Mannion, M., Kaindl, H.: Using binary strings for comparing products from software-intensive systems product lines. In: Proceedings of the 23rd International Conference on Enterprise Information Systems, ICEIS 2021, 26–28 April, vol. 2, pp. 257–266. SCITEPRESS (2021). ISBN 978-989-758-509-8
3. Kaindl, H., Mannion, M.: A feature-similarity model for product line engineering. In: Schaefer, I., Stamelos, I. (eds.) ICSR 2015. LNCS, vol. 8919, pp. 34–41. Springer, Cham (2014). https://doi.org/10.1007/978-3-319-14130-5_3
4. Mannion, M., Kaindl, H.: Using similarity metrics for mining variability from software repositories. In: 2014 Proceedings of the 18th International Software Product Lines Conference - Companion Volume for Workshop, Tools and Demo Papers, SPLC 2014, Florence, Italy, 15–19 September, pp. 32–35. ACM (2014). ISBN 978-1-4503-2739-8

Unleash the Power of Engineering Questions

Neil B. Harrison[1,2](✉) and Ademar Aguiar[1,2]

[1] Faculty of Engineering, INESC TEC, University of Porto, Porto, Portugal
neil.harrison@uvu.edu, ademar.aguiar@fe.up.pt
[2] Utah Valley University, Orem, UT 84058, USA

Abstract. This tutorial delves into the transformative power of asking effective questions in engineering information systems. We explore how crafting well-defined questions in both the problem space (what issue are we addressing?) and the solution space (how will we approach it?) is paramount for success. The session will unveil the intricate relationship between these questions – how the "what" shapes the "how" and vice versa. We move beyond the fear of asking "naive" questions, demonstrating how these can spark innovation and reveal hidden assumptions. By the end, attendees will have a powerful and easy-to-use technique that removes the fear from questions.

Keywords: Engineering Information Systems · Software Architecture · Software Design · Systems Engineering

1 Background: System Design Questions

Early design and engineering of a system is by nature dominated by unknowns. During the conceptual phase of engineering, the system being envisioned is typically fuzzy: end users and other stakeholders often do not know exactly what they want, and even might have only a vague picture of their desires. Of course, the system being designed is completely uncertain. The process of engineering a system includes asking many questions. The answers to those questions shape the design of the system, and thus are an essential part of the design process.

Unknowns – questions about systems – naturally fall into two areas, the problem space, and the solution space. The problem space concerns the capabilities of the system as seen by stakeholders; often referred to as "requirements" or "specifications." On the other hand, the solution space is about how the system is designed and implemented. They are not independent; answers to problem space questions may be impact solution space questions and vice versa.

Problem space techniques such as requirements engineering can elicit information about how the system should behave. Similarly, quality attribute analysis techniques focus on understanding the quality attribute requirements of a system. There are also techniques for designing systems, such as object-oriented design for software systems. While these are useful, they focus on either the problem space or the solution space. But during system design, questions are intermingled [1]. For example, a design question can easily prompt a corresponding quality attribute question. This intermingling of questions demands an integrated approach to managing them.

J. Araújo et al. (Eds.): RCIS 2024, LNBIP 514, pp. 166–167, 2024.
https://doi.org/10.1007/978-3-031-59468-7

Questions that arise during the engineering of an engineering system have differing times in which they must be answered. In many cases, this is because some questions depend on the answers to other questions before they can get answered. Some questions can block progress in the design. In a survey we conducted of system architects, several admitted that if such questions are not answered immediately, they occasionally make educated guesses about the answers to be able to make progress. This creates the potential for costly rework if the assumptions about the questions were wrong.

2 An Integrated Questioning Technique

We have developed a simple, easy-to-use technique for eliciting and managing questions that arise during system engineering. It is based on several techniques used in requirements analysis, quality attribute analysis [2], and architecture reviews [3]. Questions are organized by the topics relevant to the current project, in both the problem space and the solution space.

Questions are also organized by when they must be answered. A fixed time scale is not necessary, but rather can be represented by dependencies of questions on earlier questions. The time, as well as the logical organization, are easy to visualize.

The technique includes a method for eliciting the questions. However, this particular question elicitation exercise is not required. Other techniques may be used, and some organizations may not wish to formally elicit questions at all. Instead, they may simply manage questions as they arise during the course of engineering design. In any case, questions are added, removed, and updated at any time.

It is easy to visualize the questions: to see the groups of question topics, the time dependencies, as well as the general state of questions. It is also low-tech in that no computer hardware or software is needed.

We have validated this technique in a controlled experiment in which eleven teams participated in an exercise of question elicitation and organization. The experiment shows it to be effective at eliciting and organizing questions by topic as well as time.

Disclosure of Interests. The authors have no competing interests to declare that are relevant to the content of this article.

References

1. Harrison, N., Avgeriou, P., and Zdun, U.: Focus group report: capturing architectural knowledge with architectural patterns. In: Proceedings of the 11th European Conference on Pattern Languages of Programs (EuroPLoP 2006), pp. 691–695. Springer, Irsee Germany (2006)
2. Barbacci, M., et al.: Quality attribute workshops (QAWs) (2003)
3. Maranzano, J., et al.: Architecture reviews: practice and experience. IEEE Softw. **2**(2), 34–43 (2005)

Author Index

J. Araújo et al. (Eds.): RCIS 2024, LNBIP 514, pp. 169–171, 2024.
https://doi.org/10.1007/978-3-031-59468-7

Printed in the United States
by Baker & Taylor Publisher Services